Diet Selection

*An Interdisciplinary Approach
to Foraging Behaviour*

EDITED BY R.N. HUGHES
Professor in Zoology at the
School of Biological Sciences
University of Wales
Bangor

OXFORD
BLACKWELL SCIENTIFIC PUBLICATIONS
LONDON EDINBURGH BOSTON
MELBOURNE PARIS BERLIN VIENNA

© 1993 by
Blackwell Scientific Publications
Editorial Offices:
Osney Mead, Oxford OX2 0EL
25 John Street, London WC1N 2BL
23 Ainslie Place, Edinburgh EH3 6AJ
238 Main Street, Cambridge
  Massachusetts 02142, USA
54 University Street, Carlton
  Victoria 3053, Australia

Other Editorial Offices:
Librairie Arnette SA
2, rue Casimir-Delavigne
75006 Paris
France

Blackwell Wissenschafts-Verlag
Meinekestrasse 4
D-1000 Berlin 15
Germany

Blackwell MZV
Feldgasse 13
A-1238 Wien
Austria

First published 1993

Set by D & N Publishing
The Old Surgery, Lambourn, Berkshire
Phototypeset by Fido Imagesetting,
Witney, Oxon
Printed and bound in Great Britain
at the University Press, Cambridge

DISTRIBUTORS

Marston Book Services Ltd
PO Box 87
Oxford OX2 0DT
(*Orders*: Tel: 0865 791155
        Fax: 0865 791927
        Telex: 837515)

USA
Blackwell Scientific Publications, Inc.
238 Main Street,
Cambridge, MA 02142
(*Orders*: Tel: 800 759–6102
            617 876–7000)

Canada
Oxford University Press
70 Wynford Drive
Don Mills
Ontario M3C 1J9
(*Orders*: Tel: 416 441–2941)

Australia
Blackwell Scientific Publications
Pty Ltd
54 University Street
Carlton, Victoria 3053
(*Orders:* Tel: 03 347–5552)

A catalogue record for this title
is available from the British Library

ISBN 0–632–03559–5

Library of Congress
Cataloging-in-Publication Data

Diet selection: an interdisciplinary approach to
    foraging behaviour/edited by R.N. Hughes.
        p.   cm.
    Includes bibliographical references
    and index.
    ISBN 0–632–03559–5
    1. Animals–Food.   2. Food preferences.
    I. Hughes, R.N.
    QL756.5.D54   1993
    591.53—dc20

Contents

iii

List of Contributors

RICHARD P. CINCOTTA *Range Science Department, Utah State University, Logan, Utah 84322-5230, USA*

WILLIAM R. DEMOTT *Department of Biological Sciences and Crooked Lake Biological Station, Indiana University-Purdue University, Fort Wayne, Indiana 46805, USA*

IAIN J. GORDON *Macaulay Land Use Research Institute, Craigie Buckler, Aberdeen AB9 2QJ, UK*

ROGER N. HUGHES *School of Biological Sciences, University of Wales, Bangor, Gwynedd LL57 2UW, UK*

ALASDAIR I. HOUSTON *NERC Unit of Behavioural Ecology, Department of Zoology, University of Oxford, South Parks Road, Oxford OX1 3PS, UK*

ANDREW W. ILLIUS *Institute of Cell, Animal and Population Biology, Ashworth Laboratories, University of Edinburgh, West Mains Road, Edinburgh EH9 3JT, UK*

PETER A. JUMARS *School of Oceanography, WB-10, University of Washington, Seattle, Washington 98195, USA*

DEBORAH L. PENRY *Department of Integrative Biology, University of California at Berkeley, Berkeley, California 94720, USA*

CATHERINE M.S. PLOWRIGHT *Department of Psychology, University of Ottawa, Ottawa, Ontario K1N 6N5, Canada*

FREDERICK D. PROVENZA *Range Science Department, Utah State University, Logan, Utah 84322-5230, USA*

PAMELA J. REID *Department of Psychology, University of Massachusetts, Massachusetts, Amherst, Massachusetts 01003, USA*

SARA J. SHETTLEWORTH *Department of Psychology, University of Toronto, Toronto, Ontario M5S 1A1, Canada*

ANDREW SIH *T.H. Morgan School of Biological Sciences, 101 Morgan Building, University of Kentucky, Lexington, Kentucky 40506-0225, USA*

Preface

All animals feed selectively. The rumen of a cow will contain a biased sample of vegetation, the size frequency of particles ingested by a marine worm will differ from that of the native sediment and a predator will prefer certain prey. To the dietitian, food selection is interesting for its nutritional consequences, to the behavioural ecologist it poses questions about decision rules, to the population ecologist it bears upon dynamical stability, to the community ecologist it is a factor influencing species diversity and to the evolutionary ecologist it is an important ingredient of niche segregation.

Such different points of view have much to offer one another, provided there is common ground. Optimal foraging theory has played a large part in preparing this ground, as well as having been a major force in behavioural ecology. Its simplistic approach has brought the study of foraging behaviour, particularly diet selection, to a point where inherent complexities, including physiological, nutritional, psychological, morphological and ecological factors can begin to be addressed in a coherent fashion. While not intending to write yet another exposition on optimal foraging theory, we have drawn upon its applications and limitations to demonstrate the great potential for development of diet selection as an interdisciplinary subject.

In the introductory chapter, Hughes describes the success of optimal foraging theory as applied by anthropologists to hunter-gatherers, then illustrates its limitations with reference to the behaviourally inflexible foraging patterns of ants. Houston in Chapter 2 shows how an animal's state, for example its level of energy reserves or its relationships with other animals while foraging, can be modelled by dynamic programming. Penry, in Chapter 3, discusses the importance of choosing the appropriate type of model when dealing with digestive kinetics.

Moving from physiological to behavioural considerations, Shettleworth, Reid and Plowright (Chapter 4) examine the psychology of diet selection, an important aspect not treated by basic optimal foraging theory. In a similar vein, Provenza and Cincotta (Chapter 5) stress the importance of learning in the rapid adjustment of selective feeding to varying food quality. In Chapter 6, DeMott shows how optimal foraging theory successfully predicts the selective feeding of zooplankton, as long as hunger is taken into account. He emphasizes that even the simplest consumers can exhibit relatively complex behaviours and even the smallest prey are attacked by predators that can ingest them individually, rather than filtering them in bulk. Jumars (Chapter 7) finds that deposit feeders, for example sediment-swallowing worms, are obliged to maintain a rapid throughput of nutritionally dilute ingesta. Here, selective feeding based on particle size is achieved by physical, non-behavioural mechanisms.

In contrast to deposit feeders, terrestrial herbivores using fermentation have slow throughput rates. Their food is refractive but ultimately digestible and selective feeding enhances the quality of the brew. Diet selection by these animals, however, cannot be fully understood without reference to constraints associated with body size and perceptual faculties (Illius & Gordon, Chapter 8).

Animals seldom forage in isolation, but often face problems of competition and the risk of being eaten themselves. In Chapter 9, Sih explores some of the effects that these factors might have on diet selection. In so doing, he reveals an exciting link between behavioural, population and community ecology.

R.N. Hughes

Acknowledgements

We thank Susan Sternberg of Blackwell Scientific Publications for encouragement throughout the preparation of this book. Specific acknowledgements are as follows.

Chapter 1: Critical comments: Malcolm Cherrett, Alasdair Houston, Andrew Illius and Kim Hill. Malcolm Cherrett, Kim Hill, Paul Rozin, Eric Alden Smith and Bruce Winterhalder generously supplied reprints.

Chapter 2: Critical comments: Roger Hughes, John McNamara, Wayne Thompson, Wolfgang Weisser. W. Thompson produced the figures. A. Houston was supported by the NERC.

Chapter 3: Critical comments: R. McN. Alexander, P.A. Jumars, M.R. Roman and D.P. Weston. Work was supported in part by a grant from the Office of Naval Research (N00014-90-J-1089) and by a postdoctoral fellowship from the University of Maryland, Horn Point Environmental Laboratory. Contribution #2255 from the University of Maryland.

Chapter 4: Critical comments: Alex Kacelnick. S.J. Shettleworth and C.M.S. Plowright were supported by operating grants from the Natural Sciences and Engineering Research Council of Canada. P.J. Reid was supported in part by a postgraduate scholarship from NSERC.

Chapter 5: Critical comments: Carl Cheney, Andrew Illius and Justin Lynch. F.D. Provenza was supported in part by a grant from Co-operative Research Service.

Chapter 6: Critical comments: Hank Vanderploeg. W.R. DeMott was supported in part by NSF grant BSR 9006770; his work on *Eudiaptomus* was undertaken at the Max Planck Institute of Limnology.

Chapter 7: Critical comments: B. Hentschel, L. Mayer, D. Penry, C. Plante and R. Zimmer-Faust. P.A. Jumars was supported by NSF grant OCE-89-06425 and ONR grant N00014-90-J-1078.

Chapter 8: Critical comments: Frederick Provenza. A. Illius was supported in part by a grant from the NERC.

Chapter 9: Critical comments: J.F. Gilliam. A. Sih was supported by NSF grants 88-18028 and 90-20870. Many of the ideas for this chapter originated at a NATO conference in Wales, which resulted in the volume *Behavioural Mechanisms of Food Selection*, edited by R.N. Hughes, and subsequently took shape through helpful discussions with M.-S. Baltus and L. Sih.

1: Introduction

ROGER N. HUGHES

BASIC OPTIMAL DIET THEORY

Feeding is of pressing importance to most animals for most of the time. It balances the energy budget and determines nutrient status, exercising a primary influence on scope for further activity, growth, reproduction and correlated aspects of fitness. Feeding is a major basis for taxonomic diversity, simplistically caricatured in the concept of trophic groups. Trophic interactions influence population dynamics and community structure. Together with competition, such interactions shape ecological niches and, on an evolutionary time scale, help to drive adaptive radiation. Not surprisingly, therefore, interest in the biology of feeding has a long pedigree among physiologists, dietitians, psychologists, ethologists, ecologists and evolutionary biologists alike. Points of view have differed, but each is relevant to the others. For mutual interest to reach fruition some unification is required and, to a significant extent, this has been provided by optimal foraging theory.

Optimal foraging theory probably first became recognizable as such following publication of two seminal papers by MacArthur & Pianka (1966) and Emlen (1966). These authors wanted to predict how natural selection moulds patterns of foraging behaviour. They assumed that an animal would promote its fitness by foraging in ways that maximize the net rate of energy gain (E/T). This energy-maximization premise underpins the whole of optimal foraging theory, in its application to searching behaviour, exploitation of food sources and selection among alternative food items (Stephens & Krebs 1986). The subset devoted to food selection, optimal diet theory, has been built upon a simple model of polyphyletic origin.

The basic optimal diet model makes many simplifying assumptions, the most fundamental of which are that a forager can evaluate the profitability, in terms of yield per unit handling time, of each food type

1

encountered and rank this relative to the profitabilities of other types; that the forager can estimate and remember the average profitability of food types encountered; that the forager can measure encounter rates with different food types; and that the forager uses all this information to decide which encountered items to accept and which to reject. The two strongest predictions of the model are that a forager should always accept the most profitable food type and that it should accept successively less profitable types only when encounter rates with higher-ranking types fall below critical levels. The diet therefore should expand and contract according to the quality and availability of alternative foods.

Human foraging behaviour

In applying the basic optimal diet model, ideally one would like to know about the sensory capabilities of the forager, its capacity for remembering and using information gained during foraging, and its motivation. The more thoroughly such factors are known, the more convincing will be any comparison between observation and prediction. What better chance of achieving this knowledge than with our own species? Some of the most convincing field tests of basic optimal foraging theory, including diet selection and patch use, have been made by anthropologists observing the foraging behaviour of hunter-gatherers, among whom they lived and established a rapport (Winterhalder 1981; Hill & Hawkes 1983; Smith 1991).

There has been good agreement between observation and prediction. Winterhalder's (1981) pioneering study of Cree hunter-gatherers made ingenious use of changing technology, through its effect on searching speed and prey-encounter rates. Historical records and verbal reports showed that in progressing from paddled canoes, through motorized canoes to snowmobiles, the Cree dramatically increased their searching efficiency and, as predicted, hunted more selectively, eventually ignoring less profitable, usually smaller species that once had been acceptable quarry.

Similar use of technological contrast was made by Hill & Hawkes (1983) in their study of Ache hunter-gatherers in the forests of eastern Paraguay. Sometimes hunters used shotguns instead of the traditional bow and arrows, considerably increasing attack efficiency and hence the average return rate while hunting. From measurements of average return rate and prey profitabilities (based only on handling during the hunt), diet theory predicted that shotgun hunters should ignore birds weighing less than about 1 kg, whereas bow and arrow hunters should ignore only those weighing less than about 0.4 kg. In remarkable agreement, the smallest bird shot by gun weighed 1.4 kg and by bow and arrow 0.4 kg.

Agreement between observation and prediction is all very well, but does it reflect the decision-making process envisaged by foraging theory, or is it the coincidental result of other mechanisms? Such ambiguity bedevils, to a greater or lesser extent, most observations on foraging behaviour. The problem would be removed if we knew the mind of the forager when decisions were being made. But even with fellow humans it is difficult, as Smith (1991) points out, repeatedly to interrupt the forager for interrogation about the motivation behind decisions. Nevertheless it is possible, more accurately than for any other animal, to judge from behavioural signs when a decision has been made to pursue a particular prey item, and then to interpret this decision in the context of perceived availability of prey, environmental constraints and what time–energy budget the hunter is operating within.

In this way, Smith (1991) has made fascinating analyses of Inujjuamiut hunting forays, in the Canadian Arctic. For example, one November day, two experienced men and an apprentice set off in a canoe to hunt along the coast. First, the party encountered flocks of eider duck, but choppy seas meant that aiming and retrieval would have been too difficult, so these potential prey were ignored. Soon, the party entered the calmer water of an inlet and immediately shot the next eider encountered. After some unsuccessful pursuits of more eider and ringed seals, the hunters cruised around the inlet, landing at willow thickets to search for ptarmigan, whose white plumage could be seen against the snow-free landscape. This successful pursuit of ptarmigan was interrupted by the sighting of a ringed seal, but after failing to shoot the seal the ptarmigan hunt was resumed. Sealing alternated opportunistically with ptarmigan hunting until it was time to go home. On the return trip, waterfowl were pursued if close. The 6.5 h hunt resulted in 53 ptarmigan, one small ringed seal, an eider and a merganser. Taking into account handling times incurred only during the hunt, the ringed seal yielded 30.4 MJ h^{-1}, the ptarmigan a mean of 24.5 MJ h^{-1} bird^{-1} and the waterfowl a mean of 16.4 MJ h^{-1} bird^{-1}. Several important criteria met the assumptions of the basic optimal diet model. First, hunters ranked prey types in order of preference that matched profitability; thus seals were pursued whenever there was perceived to be some chance of success, ptarmigan were pursued when there was little chance of catching seal, and waterfowl were pursued only in circumstances when preferred prey were inaccessible, and even then, only when particularly vulnerable in calm water at close range. Second, encounters with different prey types were independent, and third, pursuit of one prey item precluded pursuit of another: for example, during pursuit of the seal a flock of ptarmigan was spotted but ignored. The priority given to the pursuit of seals and the variable attention paid to ptarmigan and waterfowl, depending on the availability of preferred prey, agreed with prediction.

Because the flesh of one vertebrate is biochemically rather similar to that of another, nutrient constraints are unlikely to influence foraging decisions of the Inujjuamiut. Broader diets, however, may require the inclusion of several nutritional currencies when modelling the economics of foraging behaviour. Ache hunter-gatherers forage both for plant and animal food items. If energy-return rate were the only important criterion guiding foraging behaviour, then the Ache should concentrate entirely on palm fibre, which would yield some 11.1 kJ h spent foraging. The observed foraging-return rate, however, was only about 5.1 kJ h^{-1}, because foragers preferred to hunt animals rich in fat and protein rather than gather palm fibre, rich in carbohydrate (Hill *et al.* 1987). Social interactions, moreover, may introduce yet other classes of variables influencing foraging behaviour. Hunting for game is risky in the sense that it sometimes yields very little, but sometimes produces a bonanza that can be shared among others. Hunting success attracts women in the Ache society, so hunting, rather than gathering, is promoted among the men (Hawkes 1990).

Once social culture becomes highly developed, the complexities of human behaviour begin to overshadow the simple economics of foraging envisaged by basic models. Culture has uncoupled foraging from ingestion and has greatly extended the range of food types available, through farming, commercially organized hunting and fishing, efficient transportation and storage (Rozin & Vollmecke 1986). Even with small bands of hunter-gatherers, there are potential complications because food is transported back to base for processing, distribution, or storage. Simple optimal foraging models still may apply to the hunt itself because food is processed at the end of the day, when hunting would in any case be impossible. Moreover, processing is often carried out by non-hunters. Post-hunt processing costs therefore do not strongly influence decisions made during the hunt (Smith 1991).

From the gathering, hunting, or production of food to its ingestion, there may be several interacting levels at which 'foraging' decisions are made. At the production level, decisions are made about which food types to pursue or to cultivate. At the distribution level, decisions are made about how to stock the supermarket. At the consumer level, decisions are made about what to buy and, at the dietary level, about what to select for the meal. Production, distribution and storage of food are so highly developed among industrial nations that supermarkets throughout the world offer a similar range of foods, catering for many ethnic and sociological tastes. Can the choices made when pushing the supermarket trolley, or when reaching into the fridge for the ingredients of dinner, be understood in the context of foraging theory? Clearly, a successful theory will require the combined techniques of economics, sociology, psychology, behavioural ecology and digestive physiology.

The subtleties of human predilection for particular food types has long been of interest to psychologists. It seems reasonable to suppose (Cabanac 1971) that pleasures derived from eating certain foods reflect physiological demands. By preparing nutritionally contrasted food types with distinctive flavours Booth *et al.* (1982) showed that over a series of meals, human subjects came to prefer the flavour of high-calorie food when hungry and of low-calorie food when replete.

Nutritional qualities apart, the acceptability of a food type may change as a consequence of encounter frequency, or merely according to sociological or psychological circumstances. Human subjects tend to eat more, given a varied menu, than when offered only their favourite food (Rolls *et al.* 1986). Liking for a food type, relative to alternatives, declines once it has been eaten, reaching a minimum within about 20 min, long before any post-ingestive feedback could have occurred. On the other hand, entirely new foods are liked less before some experience of them has been gained. Liking for a novel food type, therefore, is greatest at moderate levels of exposure. Although not necessarily directly linked to the nutritional qualities of particular food types, this non-linear response in predilection may be a general strategy that allows sampling to apprise the forager of the range of foods available, while preserving a cautious approach to untested novelties, yet avoiding the nutritional impoverishment of a monotonous diet.

Timing and context, also, may strongly influence food selection (Rozin & Vollmecke 1986). Different foods are preferred depending on whether it is breakfast, lunch, or evening meal. Whether or not this has physiological significance in terms of the time–energy budget would repay further investigation.

Does the forager decide?

Non-human subjects may engage in less complex webs of decision-making, yet this should not lead us to make glib comparison with basic foraging theory. The foraging behaviour of ants is a case in point. Apparent, optimal foraging behaviour of ants is not entirely the result of decision-making. On the contrary it is heavily dependent on stereotyped responses to stimuli. In the world of the ant, this stereotype works in a way that would be disadvantageous in other animals. The key to its success lies in the vast number of workers and their total subservience to the colony, possibly engendered by the high level of kinship resulting from haplo-diploid sex determination (Hamilton 1972). Ant foraging behaviour is based upon scent trails laid by foraging workers as they return to, and sometimes as they leave the colony. Workers respond positively to the concentration of scent, leading them to switch from less strongly to more strongly marked trails whenever the two meet. Having been recruited onto a trail, the worker adds her own scent, creating a positive-feedback loop that strengthens one trail at the expense of another. This inflexible mechanism, devoid of choice, is essentially stochastic, yet under natural circumstances it produces highly adaptive foraging patterns. Computer simulations show that foraging patterns characteristic of different species can be generated using a common trail-laying and trail-following behaviour, simply by altering the density and spatial distribution of food (Goss *et al.* 1990).

The adaptive significance of a particular foraging pattern becomes clear in its natural ecological setting. For example, leaf-cutting ants, *Atta* spp., build huge underground nests in tropical rainforest. They forage for leaves with which to provision the fungus garden, deep within the nest, that ultimately supplies the ants with food. Nests can persist for 20 y, a large one requiring some 62 g of vegetation h^{-1} and perhaps accounting for as much as 0.2% of the gross productivity of the forest (Lugo *et al.* 1973). In the artificial monotony of citrus groves, such exploitation is devastating, killing trees on a large scale, but in the rain-forest, where species are diverse and well-mixed, trees are seldom killed. Moreover, the long-term success of the leaf-cutter ant depends on avoiding over-exploitation of the vegetation (Cherrett 1983; Reed & Cherrett 1990). Here, the stochastic, inflexibly programmed foraging behaviour reveals its potential value. Trees suitable for harvesting are unpredictably located, partly because of their scattered distribution among non-food species and partly because their suitability for harvesting is maximal at leaf-flush, which occurs asynchronously even within species. Suitable trees, therefore, are discovered haphazardly by ants scouting the forest and once exploitation of a particular tree has begun, the trail-recruitment mechanism concentrates foraging effort upon this food source, even when others are readily accessible nearby. Cherrett (1983) observed *A. cephalotes* to defoliate 16 specimens of *Emmotum fagifolium* within 24 d. Presumably by chance, the first to be defoliated was 97 m from the nest and the last merely 3 m away, just the opposite of what should be expected on economic grounds. In another case, Cherrett (1983) examined a pair of mango trees, only 2.5 m apart, for signs of previous exploitation by leaf-cutting ants foraging from a nest 95 m away. The mango retains its leaves for several years, so by counting back from the nodes of branches and looking for the unmistakable damage inflicted by leaf cutters, he could estimate the incidence of attack over previous growing seasons. In the two immediately previous seasons, neither tree was attacked, in the third and fourth only tree A, in the fifth and sixth both trees, and in the seventh again only tree A. Such erratic exploitation of trees, closely paired relative to the ambit of the foraging ant, can best be explained by chance discovery and strict recruitment to one source at a time. The long-term result of such hit-and-miss foraging is that over-exploitation of resources is avoided. This, together with the certainty that somewhere within the diverse forest there will always be a tree suitable for harvesting, optimizes fitness of the colony, which because of its social and breeding structure, may be regarded as the unit of selection.

Other types of ant may use the same, inflexibly programmed, trail-recruitment system, but in contrast to the leaf cutters, the mechanism in this case optimizes the economics of foraging. When given alternative routes to a source of food, the Argentine ant, *Iridomyrmex humilis*, always converges on the shortest (Deneubourg *et al.* 1990). Workers lay scent on the outward and return journey and initially workers may use any route between nest and food. Since all the ants move at roughly the same speed and a more or less constant number of ants leave the nest or food source per unit time, the passage frequency along the shortest route exceeds that along the others, when routes are used randomly. The shortest route therefore gets marked more frequently and the increased concentration of scent attracts ants, which no longer embark randomly down alternative routes. This minimizes harvesting distance, in concordance with the energy-maximization premise, but it is inflexible and unable to cope with short-term changes in circumstance. Pasteels *et al.* (1987) gave another species, *Lasius niger*, the choice of a richer and poorer supply of sucrose. Because ants are stimulated to release more scent when returning from a richer source of food, the trail leading to the more concentrated sucrose solution soon became more strongly marked and recruited most of the ants. When the solutions were transposed, however, the established trail 'trapped' the foragers, compelling them to continue visiting the same location, even though it had become the poorer source of food. Under natural circumstances, the quality of food sources would not change so rapidly. Attenuation of trail-marking, together with activity of the large number of scouts, would enable the colony to track changes among food sources.

The stereotyped behaviour described above determines the foraging patterns of workers susceptible to recruitment to food sources. Decisions may already have been made for them, however, by scouts that first discover the food sources and, by laying scent, entrain recruitment. If this is true, then the stereotyped behaviour of hoards of recruits greatly amplifies the consequences of foraging decisions made by a relatively few scouts (Cherrett 1983). Clearly, the decision-making capabilities of scouts and the potential for recruits to abort if a food source proves to be inadequate, deserve investigation.

A MORE COMPREHENSIVE APPROACH

These case histories of humans and ants illustrate both the potential and limitation of basic optimal foraging theory as an explanatory tool. Its horizon is now expanding to include principles of physiology, functional anatomy, ethology, psychology, life-history and population ecology. So the theory is evolving and merging into other conceptual frameworks, and probably will lose its identity as such. The important point is that foraging behaviour has become a central issue at the interface of several major biological disciplines. Particular aspects of this exciting development are addressed in the following chapters.

REFERENCES

Booth D.A., Mather P. & Fuller J. (1982) Starch content of ordinary foods associatively conditions human appetite and satiation, indexed by intake and eating pleasantness of starch-paired flavors. *Appetite*, **3**, 163–84.

Cabanac M. (1971) Physiological role of pleasure. *Science*, **173**, 1103–7.

Cherrett J.M. (1983) Resource conservation by the leaf-cutting ant *Atta cephalotes* in tropical rain forest. In *Tropical Rain Forest: ecology and management* (ed. by S.L. Sutton,

T.C. Whitmore & A.C. Chadwick), pp. 253–63. Blackwell Scientific Publications, Oxford.

Deneubourg J.L., Aron S., Goss S. & Pasteels J.M. (1990) The self-organizing exploratory pattern of the Argentine ant. *J. Ins. Behav.* **3**, 159–68.

Emlen J.M. (1966) The role of time and energy in food preference. *Am. Nat.* **100**, 611–17.

Goss S., Beckers J.L., Deneubourg J.L., Aron S. & Pasteels J.M. (1990) How trail laying and trail following can solve foraging problems for ant colonies. In *Behavioral Mechanisms of Food Selection,* (ed. by R.N. Hughes), *NATO ASI series, vol. G20*, pp. 661–78. Springer Verlag, Berlin.

Hamilton W.D. (1972) Altruism and related phenomena, mainly in social insects. *Ann. Rev. Ecol. Syst.* **3**, 193–232.

Hawkes K. (1990) Why do men hunt? Benefits for risky choices. In *Risk and Uncertainty in Tribal and Peasant Economies* (ed. by E.A. Cashdan), pp. 145–66. Westview Press, Boulder, CO.

Hill K. & Hawkes K. (1983) Neotropical hunting among the Ache of eastern Paraguay. In *Adaptive Responses of Native Amazonians* (ed. by R. Hames & W. Vickers), pp. 139–88. Academic Press, New York.

Hill K., Hawkes K., Kaplan H. & Hurtado A. (1987) Foraging decisions among the Ache: new data and analysis. *Ethol. Sociobiol.* **8**, 1–36.

Lugo A.E., Farnworth E.G., Pool D., Jerez P. & Kaufman G. (1973) The impact of the leaf-cutter ant *Atta colombica* on the energy flow of a tropical wet forest. *Ecology*, **54**, 1292–1301.

MacArthur R.H. & Pianka E. (1966) On the optimal use of a patchy environment. *Am. Nat.* **100**, 603–9.

Pasteels J.M., Deneubourg J.L. & Goss S. (1987) Self-organization mechanisms in ant societies (I): trail recruitment to newly discovered food sources. In *From Individual to Collective Behavior in Social Insects: Les Trielles workshop* (ed. by J.M. Pasteels & J.L. Deneubourg), pp. 155–75. Birkhauser, Basel.

Reed J. & Cherrett J.M. (1990) Foraging strategies and vegetation exploitation in the leaf-cutting ant *Atta cephalotes* (L.) – a preliminary simulation model. In *Applied Myrmecology, a World Perspective* (ed. by R.K. Vander Meer, K. Jaffe & A. Cedeno), pp. 355–66. Westview Press, Boulder, CO.

Rolls B.J., Hetherington M., Burlet V. & Van Duijvenvoorde P.M. (1986) Changing hedonic response to foods during and after a meal. In *Interaction of the Chemical Senses with Nutrition* (ed. by M.A. Kare & J.G. Brand), pp. 247–268. Academic Press, Orlando.

Rozin P. & Vollmecke T.A. (1986) Food likes and dislikes. *Ann. Rev. Nutr.* **6**, 433–56.

Smith E.A. (1991) *Inujjuamiut Foraging Strategies: Evolutionary Ecology of an Arctic Hunting Economy.* Aldine de Gruyter, Hawthorn, New York.

Stephens D.W. & Krebs J.R. (1986) *Foraging Theory.* Princeton University Press, Princeton, N.J.

Winterhalder B. (1981) Foraging strategies in the boreal environment: an analysis of Cree hunting and gathering. In *Hunter-Gatherer Foraging Strategies* (ed. by E.A. Smith & B. Winterhalder), pp. 13–35. Chicago University Press, Chicago.

2: The Importance of State

ALASDAIR I. HOUSTON

INTRODUCTION

The maximization of the net rate of energetic gain provides a simple approach to foraging in general and to diet choice in particular. This 'rate maximization' approach has been widely used – see Stephens and Krebs (1986) for a review. If an animal is maximizing its rate of gain, then in a given environment it will always forage in the same way. There are, however, many contexts in which foraging behaviour either changes as a function of the duration for which the animal has been foraging, or depends on various aspects of the animal and its environment. For example, when a rat that has been deprived of food is given the opportunity to eat, it starts by eating at a high rate, but the rate declines as the cumulative intake increases (McCleery 1977). Such satiation effects are widespread (e.g. Bousfield 1933; Nuutinen & Ranta 1986). The way in which a starling feeds its young depends on the behaviour of its mate and the number of young in the brood (Wright & Cuthill 1990). The rate at which a male pied flycatcher feeds his mate during incubation depends on her nutritional state (Smith *et al.* 1989). The foraging behaviour of bumblebees depends on the colony's stores of energy (Cartar & Dill 1990), and the foraging option chosen by juncos depends on their energy budget (e.g. Caraco *et al.* 1990). The physiology and behaviour of ectotherms depends on their body temperature. All these examples can be encompassed by a characterization of an animal in terms of its state. The idea of state was developed in system theory. The state of a system can be thought of as the information that we need if we are to predict the system's behaviour. The application of this concept to animal behaviour is discussed by McFarland (1971), Sibly and McFarland (1974) and Metz (1974, 1981). From the examples just given, it can be seen that an animal's state might include internal variables such as its energy balance and external variables that characterize its environment, including other animals.

10

An animal's behaviour depends on its state. The behaviour in turn changes the state, giving rise to a sequence of behavioural actions that depends on time. If natural selection has acted on sequences of behaviour so as to maximize an animal's fitness, then a realistic optimality model must consider the whole sequence. The action taken at any particular time cannot be considered in isolation because the contribution that the action makes to fitness may depend on future actions (McNamara & Houston 1986). The appropriate mathematical theory is known as dynamic optimization. The work of Sibly and McFarland (1976) and Heller and Milinski (1979) assumed a deterministic relationship between behaviour and the resulting change in the animal's state. In many cases, however, the uncertainties inherent in an animal's environment will result in a stochastic relationship between behaviour and changes in state. As a simple example, an animal that searches for food may have only a certain probability of finding it. Working independently, McNamara and Houston (1986) and Mangel and Clark (1986) have advocated a general approach to such cases based on stochastic dynamic programming. (See also Mangel & Clark 1988; Houston & McNamara 1988 and Clark 1991.) This general approach makes it possible to explore various features that are ignored by rate maximization, such as the fact that food may have consequences for aspects of the animal's state other than its energy balance and the fact that, in order to forage, an animal may have to put itself at risk from predators. Using an explicit model of the animal's environment, the approach can be used to find the relationship between energetic gain and fitness.

Rate maximization is a very special case in which a unit of energy always makes the same contribution to fitness, regardless of the animal's state. This chapter explores the tension between the simple and often explicit solutions that emerge from rate maximization and the general but often numeric solutions of the dynamic programming equations. For a comprehensive introduction to dynamic programming in the context of animal behaviour, see Mangel and Clark (1988).

A GENERAL FRAMEWORK

State variables

A complete description of an individual would require a large number of state variables. In most circumstances, however, it will be possible to concentrate on just a few crucial components of the state. Here, we are concerned with foraging behaviour, so appropriate state variables might be the animal's level of energy reserves and the contents of its digestive system (e.g. McNamara & Houston 1982; Burrows & Hughes 1991). In general a single state variable will be denoted by x and a vector of variables by $\mathbf{x}$.

An important feature of some models is that even if an animal's state remains the same, it can be advantageous for behaviour to depend explicitly on time. For example, when foraging stops at some fixed time T, then optimal foraging decisions should depend on both energy reserves and the time remaining until T (e.g. Houston & McNamara 1985a,b; McNamara & Houston 1986; Lucas 1985, 1990).

Actions

An animal's behaviour is modelled as a sequence of distinct actions taken at time $t = 0,1,2...$ Let the animal's state at time t be $\mathbf{x}$. The animal has a set of possible actions available to it at this time. This set may depend on $\mathbf{x}$. For example, a male stickleback can only fan a nest if it has built one. The animal's state at time t and the action that it performs determines the animal's state at time $t + 1$. In many cases a realistic model involves a probabilistic relationship between the action chosen and the consequences for the animal's state. For example, if an animal searches for food, it may not find any. Therefore in general the state at time t and the action chosen determine the probability distribution of the state at time $t + 1$. In this case the model is stochastic, and the state at time $t + 1$ is a random variable $X(t + 1)$.

The action that an animal performs may make a direct contribution to the animal's reproductive success as well as changing the animal's state. Models which include such a direct contribution are not considered in this chapter – see Mangel and Clark (1988) and Clark (1991) for discussion and examples.

Change in state or dynamics

Let x denote the energy reserves of a foraging animal at time t. If the animal forages between times t and $t + 1$ then it uses an amount of energy d. If the animal is sure to find an item of food, and this item has energetic content e then its state at time $t + 1$ will be $x + e - d$. If foraging is stochastic, then the distribution of energy obtained between times t and $t + 1$ must be specified. For example, assume that if the animal is not killed, then the animal finds no item, one item or two items with probability p_0, p_1 and p_2 respectively. Then p_0 is the probability that the state at time $t + 1$ is $x - d$, i.e.

$$P(X(t + 1) = x - d) = p_0$$

Similarly

$$P(X(t + 1) = x + e - d) = p_1$$

and

$$P(X(t + 1) = x + 2e - d) = p_2.$$

If the animal is killed with probability m, then the probability that it is alive with state $x - d$ is $(1 - m)p_0$. The models of McNamara (1990a,b), McNamara and Houston (1990), McNamara, Merad and Houston (1991) are based on the above equations with $d = 1$.

Optimal decisions

Within the framework just described we can model the relationship between an animal's behaviour and its reproductive success. The animal's state plays a central role in our approach. State is important in that the contribution that an action makes to reproductive success may depend on the animal's state, and the action that the animal performs may depend on its state. As a simple example, consider a foraging animal outside its reproductive season. Intuition suggests that food is worth more in terms of avoiding starvation when the animal has low rather than high reserves of energy. Thus it is likely that behaviour will depend on reserves, with an animal being prepared to accept a greater danger of predation in its search for food when its reserves are low. The framework presented allows these intuitive ideas to be given a firm formal basis.

To introduce the approach we can start by assuming that the dependence of the animal's expected future reproductive success (efrs) on its state at the next decision time is known. This relationship is represented by a function ψ defined as follows. $\psi(x, t + 1)$ is the efrs of an animal in state x at time $t + 1$. (ψ is zero if the animal is dead.) Given this function it is possible to analyse the consequences of the actions performed by an animal in state x at time t. We define $H(x, t; a)$ to be the efrs of an animal in state x at time t that performs action a. In general, the efrs has two components, an immediate contribution and a future contribution

$\psi(\mathbf{x}, t + 1)$ that depends on the state. When there is no immediate contribution the equation for H in the stochastic case is

$$H(\mathbf{x}, t; a) = E_a \{\psi(X(t + 1), t + 1)\} \qquad (2.1)$$

where E_a denotes expectation taken over the possible outcomes when action a is chosen.

The Hs provide a common currency for evaluating different actions (McNamara & Houston 1986). The best action $a\star$ is the one with the highest H, i.e.

$$\psi(\mathbf{x}, t) = H(\mathbf{x}, t; a\star) = \underset{a}{\text{maximum }} H(\mathbf{x}, t; a) \qquad (2.2)$$

where the maximization is taken over all possible actions in state $\mathbf{x}$ at time t.

McNamara and Houston (1986) defined the canonical cost of performing action a when in state $\mathbf{x}$ at time t to be:

$$c(\mathbf{x}, t; a) = H(\mathbf{x}, t; a\star) - H(\mathbf{x}, t; a).$$

Thus the canonical cost measures the loss in terms of efrs that results from performing a single suboptimal action and then following the optimal policy. McNamara and Houston argue that the canonical costs give an indication of a model's robustness. If they are small, then the optimal action is not very much better than suboptimal actions and factors that are not included in the model may determine the action that is chosen. For further discussion, see Houston, McNamara and Thompson (1992). Houston and McNamara (1986) illustrate another measure of the selective advantage of optimal behaviour. This measure involves comparing the fitness that results from following the optimal policy with the fitness that results from following a suboptimal policy.

Equations 2.1 and 2.2 show that given a knowledge of $\psi(\mathbf{x}, t + 1)$, it is possible to calculate $\psi(\mathbf{x}, t)$. But first we need to know $\psi(\mathbf{x}, t + 1)$. This function depends on the animal's future life-history, in particular on how energy makes a contribution to reproductive success. A knowledge of it may be difficult to obtain, but is a necessary ingredient of a functional explanation of behaviour. There are various ways in which ψ can be estimated (e.g. McNamara & Houston 1986; Houston 1990). One approach is to study the whole of the animal's life. A general account of such an approach from a state-dependent perspective is given by McNamara (1991) and McNamara and Houston (1992a). This approach may be hard to implement when the behaviour of interest occurs in just a short part of the life of a relatively long-lived animal. An alternative approach is to consider some period $[0, T]$ in the animal's life. Although any final time T can be used, it is advisable to choose a biologically meaningful time, because this will help in determining the relationship between the animal's state at this time and its efrs. McNamara and Houston (1986) refer to the function $R(\mathbf{x})$ that specifies this relationship as the terminal reward.

If we are considering behaviour at time $t = T - 1$, then

$$\psi(\mathbf{x}, T) = R(\mathbf{x}) \tag{2.3}$$

and so a straightforward optimization analysis can be carried out. If there are several decisions to be made before T is reached, then we can either consider a single decision at time t, after which behaviour is fixed until $T - 1$ (e.g. McNamara & Houston 1987a) or we can allow the animal to make decisions as a function of its state at all times from t until $T - 1$. Dynamic programming is a technique for finding the best decisions under such circumstances. Equation 2.3, together with Equations 2.1 and 2.2 are the basis of dynamic programming. They can be solved numerically by working backwards from T (where ψ is given by Equation 2.3) to $T - 1$, $T - 2$ etc. This process is known as backward induction. In general, behaviour becomes independent of the terminal reward as the time left until T increases. See McNamara and Houston (1986), Mangel and Clark (1986, 1988), Houston and McNamara (1988), McNamara (1990a) and Clark (1991) for further discussion.

Some idea about the general form of the terminal reward may follow from the biology of the animal under investigation. For example, if the foraging behaviour of a small bird in winter is being studied, and the period $[0, T]$ runs from dawn until dusk, then a bird with less than a certain amount of energy at dusk will not survive the night. As a first approximation, the terminal reward can be taken to be a step function, i.e.

$$R(x) = \begin{cases} 1 \text{ if } x > x_c \\ 0 \text{ if } x \le x_c \end{cases}$$

One can then investigate how sensitive the model's predictions are to changes in the form of $R(x)$. Alternatively, R can be estimated from data on the relationship between the state at T and subsequent reproductive success. Both approaches can be combined with an exploration of the sensitivity of a model's predictions to changes in R. One can also work backwards over a series of days until the terminal reward converges (see McNamara & Houston 1986, 1992b; McNamara, Houston & Krebs 1990). This process is discussed further below.

MAXIMIZING THE ENERGY OBTAINED
FROM FORAGING

The idea of relating the energy obtained over a given period to subsequent fitness was put forward by Schoener (1971). Although Schoener's paper has been very influential, it seems that more attention has been given to the idea of maximizing the rate of energetic gain than to the more fundamental notion of relating energy to fitness. Rate maximization can be seen as a special case of the above framework in which a given increase in energy always makes the same contribution to fitness (Pyke 1984; McNamara & Houston 1987a; Mangel 1992). In other words, rate maximization assumes that the terminal reward is linear. Given this form of terminal reward, the standard results of optimal foraging theory can readily be obtained.

Let the animal have reserves x at time t and assume that it forages at rate γ until time T. The terminal reward $R(x)$ at this time is of the form $R(x) = \mathrm{a}x + \mathrm{b}$. Then, as McNamara and Houston (1987a) show,

$$\psi(x, t) = \mathrm{a}(x + \gamma \, (T - t)) + \mathrm{b}.$$

The equation for ψ can be simplified without changing the nature of the results by assuming that $\mathrm{a} = 1$, in which case

$$\psi(x, t) = x - \gamma t + k,$$

where

$$k = \gamma + \mathrm{b}.$$

We now look at a choice between eating or rejecting an item of food that gives an energetic gain e and takes a time h to handle. After this choice the animal forages at rate γ until T. These choices can be compared in terms of their Hs

$$H_{\text{eat}} = x - \gamma t + e - \gamma h + k$$

$$H_{\text{reject}} = x - \gamma t + k.$$

The best action is the one with the higher H. It follows that the animal should eat the item if and only if

$$H_{\text{eat}} > H_{\text{reject}}$$

i.e. if and only if

$$e/h > \gamma$$

This is the standard condition for including a prey item in the diet that results in the maximum rate of energetic gain (e.g. Charnov 1976; Stephens & Krebs 1986). It predicts that a given type of item should either always be accepted or always rejected. McNamara and Houston (1987b) review various reasons why such all-or-none preferences might not be found.

RISK-SENSITIVE FORAGING

The basic logic of risk-sensitive foraging is illustrated by the following example, involving a single decision (cf. McNamara & Houston 1986). An animal that can choose between two foraging options has reserves x at time t. If it chooses option 1, then it is sure to have a net gain of one unit of energy. In contrast, option 2 results in the animal having a net gain of either nothing or two units of energy. Each alternative occurs with probability 0.5. How do the options compare in terms of the resulting contribution that they make to the animal's expected future reproductive success? If the animal chooses option 1, then its reserves at time $t + 1$ are $x + 1$ and its efrs is $\psi(x + 1, t + 1)$. If the animal chooses option 2, then half the time it gets nothing, so that its efrs is $\psi(x, t + 1)$. The other half of the time it gets two units of energy so its efrs is $\psi(x + 2, t + 1)$. Let us assume that ψ does not depend on time. It follows that the animal should choose the variable option if and only if

$$\frac{\psi(x + 2) + \psi(x)}{2} > \psi(x + 1)$$

A standard result known as Jensen's inequality tells us if $\psi(x)$ is an accelerating function of x, then the inequality holds and it is optimal to prefer variability, i.e. to be risk prone. The result is illustrated in Fig. 2.1. If $\psi(x)$ is linear, then the two options are equivalent, whereas if $\psi(x)$ is decelerating, then it is optimal to choose the less variable option, i.e. to be risk-averse.

In introducing the idea of risk-sensitive foraging only a single action has been considered and a dependence of efrs on energetic gain has been assumed. As is now shown, a series of actions can be analysed, calculating the dependence of efrs on energy in the process.

A dynamic model

The animal chooses between foraging options at times $t = 0, 1, 2 \ldots T - 1$. Foraging stops at $t = T$ and a terminal reward $R(x)$ at this point

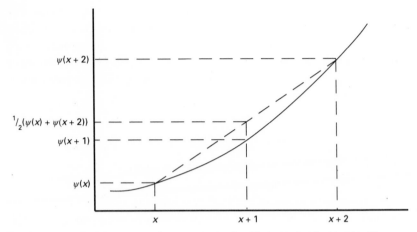

Fig. 2.1 A graphical illustration of Jensen's inequality. Under option 1, the animal has reserves $x + 1$ and fitness $\psi(x + 1)$. Under option 2, half the time the animal has reserves $x + 2$ and hence fitness $\psi(x + 2)$, whereas half the time the animal has reserves x and fitness $\psi(x)$. The resulting expected fitness is $\frac{1}{2}(\psi(x) + \psi(x + 2))$, which is greater than $\psi(x + 1)$.

characterizes the animal's efrs as a function of its level of energy reserves x. The level of reserves can never exceed the upper limit L. If at any time the level of reserves falls to zero (the lower boundary), the animal dies of starvation.

The animal can choose between two foraging options. These options are referred to as option 1 and 2. If the animal chooses option i ($i = 1,2$) then its reserves increase by one unit with probability p_i, decrease by one unit with probability q_i and stay the same with probability $1 - p_i - q_i$. These changes include metabolic expenditure, so that the mean net gain under option i is $p_i - q_i$, and its variance is $p_i + q_i - (p_i - q_i)^2$. It follows that if both options have the same mean then the option with the bigger value of $p + q$ has the bigger variance.

Houston and McNamara (1988) show how this model can be used to work backwards from a terminal reward at T to find the best action and the resulting efrs. To illustrate the process, let

$$p_1 = 0.4, \ q_1 = 0.2$$

$$p_2 = 0.5, \ q_2 = 0.3,$$

so that both options have a mean net gain of 0.2, and assume that $R(x) = 1$ if $x > x_c$ and $R(x) = 0$ if $x \leq x_c$, where x_c is the critical level of energy reserves for survival after time T. Let the upper limit L on reserves be 20 and let x_c be 9. What are the consequences of choosing either option 1 or option 2, when $x = 10$ and there is one decision left, i.e. $t = T - 1$? If the animal chooses option 1, then with probability 0.4 its level of reserves at T is 11, and with probability 0.4 its level of reserves at T is 10. Both of these levels have an efrs of 1. With probability 0.2 the level of reserves at T is 9, and this has an efrs of zero. Thus $H(10, T - 1; 1) = (0.4 + 0.4)(1.0) = 0.8$.

By an analogous argument, $H(10, T - 1; 2) = (0.5 + 0.2)(1.0) = 0.7$. By definition, $\psi(10, T - 1)$ is the maximum of these two values, and so

$$\psi(10, T - 1) = 0.8.$$

Thus when $x = 10$ and $t = T - 1$ it is optimal to choose option 1 (the low variance option) and that the resulting efrs is 0.8. The canonical cost of choosing option 2 is $0.8 - 0.7 = 0.1$. When $x = 9$ and $t = T - 1$, the terminal reward is zero unless reserves increase by one unit, in which case the terminal reward is 1. It follows that

$$H(9, T - 1; 1) = 0.4$$

and

$$H(9, T - 1; 2) = 0.5,$$

so that it is optimal to choose option 2 (the high variance option). The resulting efrs is $\psi(9, T - 1) = 0.5$.

When $x > 10$ at $T - 1$, then efrs is 1 whatever action is chosen, and when $x < 9$, efrs is zero whatever action is chosen. ψ is now known for all x at $t = T - 1$:

$$\psi(x, T - 1) = 1.0 \text{ for } 10 < x \leq L$$

$$\psi(10, T - 1) = 0.8$$

$$\psi(9, T - 1) = 0.5$$

$$\psi(x, T - 1) = 0.0 \text{ for } 0 \leq x < 9.$$

The next step is to calculate the optimal decisions as a function of state at $T - 2$. Repeating the process gives us the optimal policy as a function of state and time.

Figure 2.2 shows the policy for reserves between 1 and $L = 20$ and time between 0 and $T - 1 = 49$. We can take this time period to be a day,

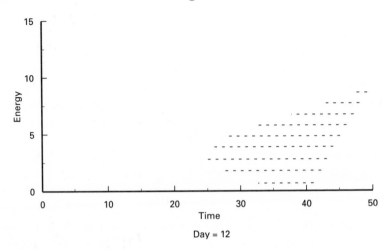

Fig. 2.2 The optimal policy over a day when an animal can choose between two options with the same mean but unequal variances. The terminal reward is a step function. The dashes represent states and times in which the more variable option should be chosen. See text for further details.

so that $t = 0$ corresponds to dawn and $t = T$ corresponds to dusk. There is a wedge-shaped region near dusk in which it is optimal to choose the more variable option. This choice is favoured because it improves the animal's chances of getting its reserves to the critical level by final time T. Avoiding starvation at the lower boundary $x = 0$ favours the less variable option. The interaction between these two effects gives rise to the wedge-shaped region. This sort of policy is typical when the means are positive and the terminal reward is a step function or an 'S'-shaped function (Houston & McNamara 1986; McNamara & Houston 1986, 1992b).

This example can be used to make some general points about dynamic programming. The terminal reward function characterises the value (in terms of fitness) of the various possible levels of energy at final time T. This final time might correspond to dusk on a given day in winter. The step function terminal reward means that all values of x below x_c are equally bad and all values of x above x_c are equally good. We could base this form of terminal reward on the probability of surviving from dusk until the following morning. If x_c is the amount of energy used during the night, then all animals with reserves at dusk below x_c die, whereas all animals with reserves above x_c survive until dawn. Working back from final time, we find a series of functions $\psi(x, T-1)$, $\psi(x, T-2)$ etc. which are no longer step functions. Some examples are given in Fig. 2.3.

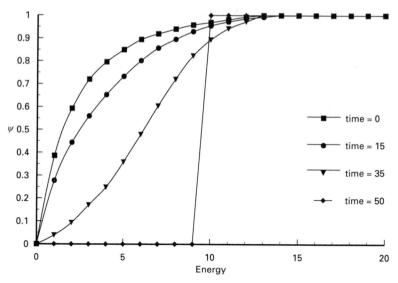

Fig. 2.3 Expected future reproductive success as a function of reserves at four times of day. At final time $T(=50)$ the terminal reward is a step function whereas at $t=35$ it is S-shaped. By time $t=15$, the fitness function is a decelerating function of reserves. Because of mortality, the fitness associated with a given level of reserves can decrease as time left until T increases. By using a normalized fitness function $\tilde{\psi}(x, t) = \psi(x, t)/\psi(L, t)$, where L is the upper limit on x, we can make the functions at different times directly comparable. The normalization changes the scale of the fitness function but not its shape. As a consequence, the optimal policy is the same for $\tilde{\psi}$ as it is for ψ. For parameter values, see text.

Now let us focus on $\psi(x, 0)$. This gives us the value of energy reserves at the start of the day. We can use this function, together with the energy used to survive the night, to find the value of energy reserves at dusk on the previous day. To express this in formal terms, we need to keep track of both time of day and day. Let the last day that we consider be day N, the next-to-last be day $N-1$, etc. and let $\psi(x, t, n)$ be the efrs of an animal with reserves x at time of day t on day n. Then

$$\psi(x, T, N) = R(x)$$

and

$$\psi(x, T, n-1) = \psi(x - x_c, 0, n).$$

If we repeat the procedure of working back over a day and then calculating the fitness function for dusk on the preceding day, we find that eventually the fitness functions at dusk converge to a particular shape that depends on the environment. This is described by McNamara and

Houston (1986, 1992b), and McNamara, Houston and Krebs (1990) and is illustrated in Fig. 2.4.

The computation of the optimal risk-sensitive policy shown in Fig. 2.2 is based on a relatively small number of possible next states when the animal forages. It is therefore surprising that the upper edge of the wedge-shaped region can be predicted by a model in which the next states are approximated by a normal distribution. This diffusion approximation was developed by McNamara and is discussed by McNamara and Houston (1992b). The same approach can also be applied when the two options differ in both mean and variance (e.g. Houston & McNamara 1985b; McNamara & Houston 1992b).

Stephens (1981) presents a simple model based on a single decision between two options with the same means but different variances. The resulting daily energy budget rule, if interpreted dynamically, gives the optimal policy when a series of decisions is made but underestimates the animal's survival probability (Houston & McNamara 1982, 1988; McNamara & Houston 1992b).

McNamara, Merad and Houston (1991) show how predictions about risk-sensitive foraging depend on the use to which an animal puts the

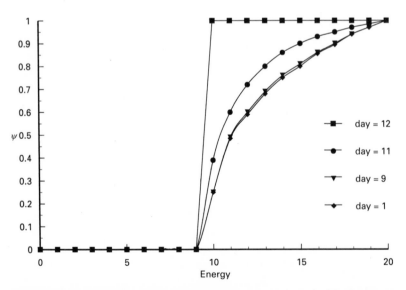

Fig. 2.4 The normalized fitness function (terminal reward) at the end of the day as we iterate back over days. The terminal reward function (TRF) on the last day ($n = 12$) is taken to be a step function. The terminal reward on day 11 is decelerating for $x \geq x_c$. After a few more days, the terminal rewards converge. See text for parameter values and further details.

energy that it obtains. They contrast a model based on maximizing long-term survival (the 'starvation' model) with one based on maximizing life-time reproductive success (the 'reproduction' model). In the reproduction model, the animal can reproduce when its reserves reach some critical level. Reproduction results in a drop in reserves, whereupon the animal can choose between two foraging options with the same mean but different variances. While foraging, there is a probability m that the animal is killed, regardless of the foraging option that it chooses. The starvation model is the same as the reproduction model except that when reserves are above the critical level, the animal rests. Optimal behaviour in the two models is strikingly different.

In the starvation model, behaviour depends on the sign of the mean gain while foraging, and does not depend on the mortality m. When the mean gain is positive, it is optimal to choose the less variable option at all levels of reserves. When the mean gain is negative, it can be optimal to choose the more variable option at low reserves and the less variable option at high reserves.

In the reproduction model, the form of the optimal policy depends on both the sign of the mean gain and on m. When the gain is positive and m is zero, it is always optimal to choose the option with the lower variance. When m is greater than zero, the optimal policy is to choose the option with the lower variance when reserves are low, and to choose the option with the higher variance when reserves are high. When the gain is negative, it is always optimal to choose the option with the higher variance.

For reviews of risk-sensitive foraging see Ellner and Real (1989), Real and Caraco (1986), McNamara and Houston (1992b) and Houston (1991).

THE TRADE-OFF BETWEEN OBTAINING FOOD AND AVOIDING PREDATORS

Many studies have shown that the danger of predation influences the foraging behaviour of animals (see Lima and Dill (1990) for a review and Chapter 9 of this book). Consider an animal in a refuge where it is safe from predators. Should the animal leave the refuge to feed in an area where it may be killed? Let the animal's state at time $t-1$ be x. Obtaining food increases x and hence increases efrs, but the animal may be killed by a predator and thus have no future reproductive success. The animal can perform one of two actions, a_1 and a_2. Action a_1 is to feed for the time interval. The animal is killed by a predator with probability m. If the animal survives, its state at time t is $x + e - d$. Action a_2 is to remain in the refuge till t. The animal is certain to survive until time t and its state at this time is $x - d$.

From the definitions,

$$H(x, t; a_1) = (1 - m)\, \psi(x + e - d, t)$$

$$H(x, t; a_2) = \psi(x - d, t).$$

It is optimal to perform action a_1 if and only if $H(x, t; a_1) > H(x, t; a_2)$ i.e. if

$$\psi(x + e - d, t) - \psi(x - d, t) > m\psi(x + e - d, t).$$

The left-hand side of this inequality is the increase in efrs that results from feeding. The right-hand side is the loss in efrs due to predation since m is the probability of death and $\psi(x + e - d, t)$ is the loss given that the animal is killed.

The condition for a_1 to be optimal can also be expressed in terms of critical level of danger m_c. It follows from the inequality that a_1 is optimal if $m < m_c$ where

$$m_c = \frac{\psi(x + e - d, t) - \psi(x - d, t).}{\psi(x + e - d, t)}$$

To illustrate this equation, assume that the animal forages from time t until final time T at a rate of γ, and that the terminal reward is $R(x) = x$.

Then as was pointed out above,

$$\psi(x, t) = x - \gamma t + k,$$

and so

$$m_c = e/(x + e - d - \gamma t + k).$$

It is instructive to note that although a given amount of energy always makes the same contribution to fitness, regardless of the animal's level of reserves, the critical level m_c is not independent of reserves. The reason for this is that the loss in efrs if the animal is killed by a predator depends on reserves. The equation shows that for given e and d, the maximum acceptable danger m_c decreases with increasing reserves x or time to go $T - t$.

Satiation

When an animal is given access to food or water after a period of depriva-
tion, it is likely to start eating or drinking at a high rate. As intake
increases, the rate of ingestion decreases (e.g. Bousfield 1933; McCleery
1977). Such 'satiation' effects have been the focus of various theoretical
investigations (e.g. Sibly & McFarland 1976; Milinski & Heller 1978;
Heller & Milinski 1979; Houston & McNamara 1989).

As a simple introduction to the question of how feeding rate should
depend on time, consider an animal with reserves $x(0)$ at time 0 that must
get its reserves to some critical level x_c at final time T. This sort of problem
was discussed in the section on risk-sensitive foraging. In this section we
simplify the problem by assuming that foraging is deterministic. If the
animal forages at an intensity u then the rate of change of reserves is

$$dx/dt = gu,$$

where g is a parameter that depends on the availability of food. The
trade-off between getting food and avoiding predators means that the
animal's mortality rate $m(u)$ is an increasing function of the foraging
intensity u.

If the animal is to survive, it must both get its reserves to x_c at T and
avoid being killed by a predator. The animal can thus adopt any feeding
trajectory $u(t)$ as a function of time that results in an energetic gain of
$x_c - x(0)$. The best policy will be the one with the lowest associated level
of predation. Houston, McNamara and Hutchinson (unpublished) show
that if $m(u)$ is an accelerating function of u, then the best policy is to
feed at a constant rate $u^\star$, where

$$u^\star = \frac{x_c - x_0}{gT}.$$

An idea of why this result holds can be obtained from considering an
alternative policy in which the animal feeds for half the time at $u^\star + \delta$
and for the other half of the time at $u^\star - \delta$. This clearly results in the
same energetic gain, but because $m(u)$ is accelerating, Jensen's inequality
(as illustrated in Fig. 2.1) means that the probability of being killed by a
predator is higher than when the animal feeds at rate $u^\star$ for the whole
period.

Houston *et al.* refer to this result as risk-spreading – the animal's best
policy is to spread the risk associated with feeding evenly throughout the
time period. They show that the result depends on various assumptions:

1   Foraging is deterministic.
2   The gain from foraging is independent of state.
3   The mortality rate is independent of state.
4   The foraging process is not subject to interruptions.

If any of these assumptions do not hold then it may be advantageous for the animal's feeding rate to depend on time, i.e. satiation effects will emerge. In this light, the decrease in $u^\star$ with time in the models of Sibly and McFarland (1976) and Heller and Milinski (1979) can be traced to their state-dependent cost terms. These costs can be thought of as mortalities, in which case assumption (3) is violated. Feeding rate also decreases with time in the 'open economy' model of Houston and McNamara (1989). In this model there is a constant probability that the foraging period will terminate. When foraging is terminated, a reward function relates future reproductive success to the animal's state at this time. All these models are deterministic. Houston *et al.* show that the general effect of adding stochasticity when the other assumptions hold is to produce a decrease in feeding rate over time.

Gilliam (1982) obtained a simple result about how a growing animal should choose between options that differ in terms of the energy that they yield and the associated predation risk. If the animal has to reach a given size in order to reproduce, and the time at which it reaches this size does not influence its reproductive success, then the animal should minimize the mortality per unit of growth. Ludwig and Rowe (1990) show that Gilliam's criterion may not result in optimal behaviour if there is a fixed time at which the critical size must be reached.

Long-term survival in a stochastic environment

In modelling satiation, the animal is assumed to be trying to move its state from one value to another, often over a limited period of time. A different approach to the trade-off between energy and predation is to assume that the animal will be in a given environment for a long period of time, and find the policy that maximizes long-term survival. McNamara (1990a) obtains some general results about the best choice of foraging option when feeding is stochastic and death may occur as a result of starvation or predation. These results are discussed and illustrated by McNamara and Houston (1990) and McNamara (1990b). As might be expected, when reserves are high, choice is largely determined by avoiding predation, whereas when reserves are low a high predation risk may be accepted if a large amount of energy can be obtained. An interesting feature of the results is that a foraging option that is not chosen very often under the optimal policy can be very important for the animal's survival. If this option is no longer available, the rate of mortality increases markedly.

SELECTIVE ADVANTAGE OF STATE-DEPENDENT BEHAVIOUR

To obtain an idea of whether natural selection is likely to favour state-dependent behaviour, we can compute the selective advantage that results from such decisions. Although the results of Houston and McNamara (1986) suggested that the selective advantage of risk-sensitive foraging was low, McNamara and Houston (1992b) found that it could sometimes give a significant benefit. In the context of food-predation trade-off there is a large advantage associated with state-dependent decisions (McNamara & Houston 1990; McNamara 1990b). These results indicate that state-dependent behaviour can substantially increase an animal's reproductive success.

More than one state variable

Although many interesting models of foraging can be based on a single state variable, there are contexts in which a realistic model must include more than one state variable. Burrows and Hughes (1991) construct a model of the foraging behaviour of the dogwhelk (*Nucella lapillus*). The state variables are the animal's level of energy reserves and the contents of its gut. The animal has the choice of staying in a safe refuge or foraging. If it forages, it can accept or reject prey items. The animal only forages when submerged by the tide, so final time T is determined by the tidal cycle. Burrows and Hughes found that the decision about whether or not to forage is strongly influenced by the contents of the animal's gut, and the decision about whether to accept a prey item depends on the time remaining until the end of the foraging period (cf. Houston & McNamara 1985a; Lucas 1990).

McNamara, Houston and Krebs (1990) investigated the hoarding behaviour of small birds in winter. The bird can store energy either as fat on its body or as items hidden in its territory. The model therefore has two state variables, one representing fat on the body and one representing hoarded items. During the hours of daylight, the bird can rest, forage and eat the items obtained, forage and hoard the items obtained, or retrieve items from the hoard and eat them. McNamara *et al.* show that there is an optimal daily routine in which items tend to be hoarded early in the day and retrieved just before dusk (see also Lucas & Walter 1991).

CONCLUSION

Rate maximization is appealing. It makes simple predictions that do not depend on state. It gives the illusion of being general, but it actually makes a specific assumption about the value of energy. Mangel (1989) points out a tension between rate maximization's assumption that the interval under study is long enough so that only the mean rate of gain is needed to compute fitness gain, and the assumption that behaviour does not change over the interval.

Rate maximization can handle some constraints (e.g. Pulliam 1975; Houston & Carbone 1992), but it cannot make predictions about risk-sensitive foraging or the trade-off between energy and predation. It also cannot deal with foraging versus other activities, e.g. singing. The general state-dependent approach can – see for example McNamara, Mace and Houston (1987).

The framework outlined here has the advantage of providing a general approach to modelling the functional significance of behaviour. It makes it possible to include constraints and stochasticity and to compare qualitatively different behaviours. Its disadvantage is that it is based on finding numerical rather than analytic solutions (see Houston & McNamara 1988; Clark 1991; Houston, McNamara & Thompson (1992) for further discussion). Any model involves simplifications. In assessing a model it is important to be aware of these. Rate maximization ignores many aspects of a forager's biology. Although it may not always be necessary to use dynamic programming, there are clearly many contexts in which rate-maximization is inadequate.

REFERENCES

Bousfield W.A. (1933) Certain quantitative aspects of the feeding behaviour of cats. *Psychonom. Sci.* 11, 263–4.
Burrows M.T. & Hughes R.N. (1991) Optimal foraging decisions by dogwhelks, *Nucella lapillus* (L): influences of mortality risk and rate-constrained digestion. *Funct. Ecol.* 5, 461–75.
Caraco T., Blanckenhorn W.U., Gregory G.M., Newman J.A., Recer G.M. & Zwicker S.M. (1990) Risk-sensitivity: ambient temperature affects foraging choice. *Anim. Behav.* 39, 338–45.

Cartar R.V. & Dill L.M. (1990) Why are bumble bees risk-sensitive foragers? *Behav. Ecol. Sociobiol.* **26**, 121–7.

Charnov E.L. (1976) Optimal foraging: Attack strategy of a mantid. *Am. Nat.* **110**, 141–51.

Clark C.W. (1991) Modeling behavioral adaptations. *Behav. Brain Sci.* **14**, 85–117.

Ellner S. & Real L.R. (1989) Optimal foraging models for stochastic environments; are we missing the point? *Comments Theor. Biol.* **1**, 129–58.

Gilliam J.F. (1982) *Foraging under mortality risk in size-structured populations.* Ph.D. Thesis, Michigan State University.

Heller R. & Milinski M. (1979) Optimal foraging of sticklebacks on swarming prey. *Anim. Behav.* **27**, 1127–41.

Houston A.I. (1990) Foraging in the context of life-history: general principles and specific models. In: *Behavioral mechanisms of food selection,* (ed. by R.N. Hughes), *NATO ASI series, vol. G 20,* pp. 23–8. Springer Verlag, Berlin.

Houston A.I. (1991) Risk-sensitive foraging theory and operant psychology. *J. Exp. Anal. Behav.* **56**, 585–9.

Houston A.I. & McNamara J.M. (1982) A sequential approach to risk-taking. *Anim. Behav.* **30**, 1260–1.

Houston A.I. & McNamara J.M. (1985a) A general theory of central-place foraging for single-prey loaders. *Theor. Popul. Biol.* **28**, 233–62.

Houston A.I. & McNamara J.M. (1985b) The choice of two prey types that minimises the probability of starvation. *Behav. Ecol. Sociobiol.* **17**, 135–41.

Houston A.I. & McNamara J.M. (1986) Evaluating the selection pressure on foraging decisions. In *Relevance of Models and Theories in Ethology* (ed. by R. Campan and R. Zayan), pp. 61–75. Toulouse.

Houston A.I. & McNamara J.M. (1988) A framework for the functional analysis of behaviour. *Behav. Brain Sci.* **11**, 117–54.

Houston A.I. & McNamara J.M. (1989) The value of food: effects of open and closed economies. *Anim. Behav.* **37**, 546–62.

Houston A.I. & Carbone C. (1992) The optimal allocation of time during the diving cycle. *Behav. Ecol.* **3**, 255–265.

Houston A.I., McNamara J.M. & Hutchinson J.M.C. (unpublished) Some general results concerning the trade-off between gaining energy and avoiding predation.

Houston A.I., McNamara J.M. & Thompson W. (1992) On the need for a sensitive analysis of optimization models. *Oikos.* **63**, 513–17.

Lima L. & Dill L.M. (1990) Behavioral decisions made under the risk of predation: a review and prospectus. *Canad. J. Zool.* **68**, 619–40.

Lucas J.R. (1985) Time constraints and diet choice: different predictions from different constraints. *Am. Nat.* **126**, 680–705.

Lucas, J.R. (1990) Food requirement and risk-sensitive foraging in shortfall minimizers. In: *Behavioral Mechanisms of Food Selection,* (ed. by R.N. Hughes), *NATO ASI series, vol. G 20,* pp. 165–84, Springer Verlag, Berlin.

Lucas J.R. & Walter L.R. (1991) When should chickadees hoard food? Theory and experimental results. *Anim. Behav.* **41**, 579–601.

Ludwig D. & Rowe L. (1990) Life-history strategies for energy gain and predator avoidance under time constraints. *Am. Nat.* **135**, 686–707.

Mangel M. (1989) Evolution of host selection in parasitoids: does the state of the parasitoid matter? *Am. Nat.* **133**, 688–705.

Mangel M. (1992) Rate maximizing and state variable theories of diet selection. *Bull. Math. Biol.* **54**, 413–22.

Mangel M. & Clark C.W. (1986) Towards a unified foraging theory. *Ecology.* **67**, 1127–38.

Mangel M. & Clark C.W. (1988) *Dynamic Modeling in Behavioral Ecology.* Princeton University Press, Princeton.

McCleery R.H. (1977) On satiation curves. *Anim. Behav.* **25**, 1005–15.

McFarland D.J. (1971). *Feedback Mechanisms in Animal Behaviour.* Academic Press, London.

McNamara J.M. (1990a) The policy which maximizes long-term survival of an animal faced with the risks of starvation and predation. *Adv. Appl. Prob.* **22**, 295–308.

McNamara J.M. (1990b) The starvation-predation trade-off and some behavioural and ecological consequences. In *Behavioral Mechanisms of Food Selection* (ed. by R.N. Hughes), *NATO ASI Series, vol. G 20*, pp. 39–58. Springer Verlag, Berlin.

McNamara J.M. (1991) Optimal life histories: a generalisation of the Perron-Frobenius theorem. *Theor. Popul. Biol.* **40**, 230–44.

McNamara J.M. & Houston A.I. (1982) Short-term behaviour and life-time fitness. In *Functional Ontogeny* (ed. by D.J. McFarland), pp. 60–87. Pitman, London.

McNamara J.M. & Houston A.I. (1986) The common currency for behavioral decisions. *Am. Nat.* **127**, 358–78.

McNamara J.M. & Houston A.I. (1987a) A general framework for understanding the effects of variability and interruptions on foraging behaviour. *Acta Biotheoretica.* **36**, 3–22.

McNamara J.M. & Houston A.I. (1987b) Partial preferences and foraging. *Anim. Behav.* **35**, 1084–99.

McNamara J.M. & Houston A.I. (1990) Starvation and predation in a patchy environment. In *Living in a Patchy Environment* (ed. by I. Swingland & B. Shorrocks), pp. 23–43. Oxford University Press, Oxford.

McNamara J.M. & Houston A.I. (1992a) State-dependent life-history theory and its implications for optimal clutch size. *Evol. Ecol.* **6**, 170–85.

McNamara J.M. & Houston A.I. (1992b) Risk-sensitive foraging – a review of the theory. *Bull. Math. Biol.* **54**, 355–78.

McNamara J.M., Houston A.I. & Krebs J.R. (1990) Why hoard? The economics of food storage in tits. *Behav. Ecol.* **1**, 12–23.

McNamara J.M., Mace R.H. & Houston A.I. (1987) Optimal daily routines of singing and foraging. *Behav. Ecol. Sociobiol.* **20**, 399–405.

McNamara J.M., Merad S. & Houston A.I. (1991) Risk-sensitive foraging for a reproducing animal. *Anim. Behav.* **41**, 787–92.

Metz J.A.J. (1974) Stochastic models for the temporal fine structure of behaviour sequences. In *Motivational Control Systems Analysis* (ed. by D.J. McFarland), pp. 5–86. Academic Press, London.

Metz J.A.J. (1981) *Mathematical representations of the dynamics of animal behaviour.* Unpublished Ph.D. thesis, University of Leiden, Leiden.

Milinski M. & Heller R. (1978) Influence of a predator on the optimal foraging behaviour of sticklebacks (*Gasterosteus aculeatus*). *Nature*, **275**, 642–4.

Nuutinen V. & Ranta E. (1986) Size-selective predation on zooplankton by the smooth newt, *Triturus vulgaris. Oikos*, **47**, 83–91.

Pulliam H.R. (1975) Diet optimization with nutrient constraints. *Am. Nat.* **109**, 765–68.

Pyke G.H. (1984) Optimal foraging theory: a critical review. *Ann. Rev. Ecol. Syst.* **15**, 523–75.

Real L.A. & Caraco T. (1986) Risk and foraging in stochastic environments: theory and evidence. *Ann. Rev. Ecol. Syst.* **17**, 371–90.

Schoener T.W. (1971) Theory of feeding strategies. *Ann. Rev. Ecol. Syst.* **2**, 369–404.

Sibly, R.M. & McFarland D.J. (1974) A state space approach to motivation. In *Motivational Control Systems Analysis* (ed. by D.J. McFarland), pp. 213–50. Academic Press, London.

Sibly R.M. & McFarland D.J. (1976) On the fitness of behaviour sequences. *Am. Nat.* **110**, 601–17.

Smith H.G., Källander H., Hultman J. & Sanzén B. (1989) Female nutritional state affects the rate of male incubation feeding in the pied flycatcher *Ficedula hypoleuca*. *Behav. Ecol. Sociobiol.* **24**, 417–20.

Stephens D.W. (1981) The logic of risk-sensitive foraging preferences. *Anim. Behav.* **29**, 628–9.

Stephens D.W. & Krebs, J.R. (1986) *Foraging Theory*. Princeton University Press, Princeton, N.J.

Wright J. & Cuthill I. (1990) Biparental care: short-term manipulation of partner contribution and brood size in the starling *Sturnus vulgaris*. *Behav. Ecol.* **1**, 116–24.

3: Digestive Constraints on Diet Selection

DEBORAH L. PENRY

INTRODUCTION

Digestion is constrained by diet selection. Less obviously, digestive capability exerts major constraints on diet choice. Diet selection is modelled most commonly as a problem in food acquisition, but it is equally a problem in food utilization. An animal realizes no gain unless it is able to digest and absorb the foods that it acquires. Furthermore, the time required for digestion of food materials is usually very much greater than the time required to obtain and ingest them (Kaiser *et al.* 1992), potentially making digestion overshadow pre-ingestive handling time as the dominant constraint in diet choice (Dade *et al.* 1990). Even more fundamentally, the animal's rate of gain in an optimality context is the rate at which digestive products are absorbed (Sibly 1981; Dade *et al.* 1990).

Optimality theory serves as the general framework for this analysis of interactions between diet and digestion. To borrow the language of chemical process and reactor design, the premise here is that an animal 'operates' to maximize its net rate of gain of energy and nutrients from foraging and digestion, subject to constraints. Digestion is explicitly considered to comprise digestive reactions and absorption across the gut wall. The goal is not to identify an optimal foraging strategy or an optimal digestive strategy (Penry & Jumars 1987) or to predict an optimal diet or gut throughput time (Dade *et al.* 1990). Instead it is to consider the constraints, the bounds within which an optimal strategy, an optimal diet, or an optimal gut throughput time must fall.

Constraints upon a process define the mathematical model of that process. A brief examination of how diet choice can be constrained by an animal's digestive capabilities and how its digestive capabilities can be constrained by its diet provides the context for comparison of compartmental models (e.g. Brandt & Thacker 1958; Ellis *et al.* 1979; Hughes & Matis 1984; France *et al.* 1990) and reactor models (Penry & Jumars

1986, 1987; Dade *et al.* 1990). Compartmental models of digestion are derived empirically, while gut-reactor models are derived theoretically. In general, compartmental models of digestion have been developed in order to compare diets (e.g. Balch 1950; Meterns & Ely 1979; Van Soest 1982; Aitchison *et al.* 1986) while gut-reactor models based on chemical reactor theory have been developed in order to compare digestive strategies (Penry & Jumars 1986, 1987; Hume 1989; Jumars & Penry 1989; Dade *et al.* 1990; Penry & Jumars 1990; Alexander 1991), though either approach could be used for either purpose.

Traditionally, diet choice and diet quality have been described and quantified in terms of digestibility, the fraction of ingested food that is digested, or assimilation efficiency, the fraction of ingested food that is incorporated into body tissue (note that the terms 'assimilation efficiency' and 'absorption efficiency' are sometimes used incorrectly as synonyms). When one attempts, however, to formulate a mathematical model of a digestive process it becomes very apparent that digestibility and assimilation efficiency tell only 'fractions' of the story. Model development demonstrates that diets should be compared in terms of an animal's net rate of gain of energy or limiting nutrients. In essence, at the very least, modelling digestion shows that measurements of digestibility and assimilation efficiency are meaningless without concurrent measurements of gut throughput time.

INTERPLAY BETWEEN DIET AND DIGESTION

Diet and digestive processes are linked so tightly that often one cannot be analyzed and understood without reference to the other. Examples are varied and numerous. In mice intestinal transporters for glucose and proline are induced or suppressed by the levels of their substrates in the diet (Karasov *et al.* 1983). Starlings lack the enzyme sucrase and, as a result, select fruits with low sucrose:glucose ratios (Martinez del Rio & Stevens 1989). In marine suspension-feeding copepods, digestive enzyme complements and activities can vary in response to food availability and chemical composition of the diet (Harris *et al.* 1986; Hassett & Landry 1990). The activity of laminarinase, for example, was observed to decline to about 50% of its initial activity when laminarin was absent from the diet of *Calanus pacificus* (Hassett & Landry 1990). Green turtles (*Chelonia mydas*) feed on seagrasses and macroalgae but generally specialize on one or the other in areas where both are abundant (Mortimer 1982). Green turtles are hindgut fermenters. They appear to develop a gut microflora specific to a seagrass diet or a macroalgal diet and thus to digest the other with much lower efficiency (Bjorndal *et al.* 1991). These examples demonstrate that interactions between diet and digestion are 'a complex interplay between physiology and ecology that involves variously flexible constraints on adaptation at various time scales' (Karasov & Diamond 1988). As complex and as varied as these interactions are, they can be characterized in terms of three relatively simple dichotomies.

Characterization of digestive constraints on diet choice

Stephens and Krebs (1986) classify constraints as either intrinsic or extrinsic. Intrinsic constraints are set by an animal's physiological abilities or tolerances, and extrinsic constraints are imposed on an animal by its environment. Digestive physiology comprises a set of intrinsic constraints on diet choice. Intrinsic constraints can be modified by extrinsic factors, and this potential for modification is especially important in interactions between diet and digestion.

Since optimal foraging and digestion theories focus on the individual, the time scales of adaptation considered in this analysis are those relevant to an individual. In this context it is possible to distinguish between constraints that result from physiological limitations, i.e. intrinsic constraints that vary only on evolutionary time scales and thus are fixed during the lifetime of an individual, and constraints that result from physiological acclimation, i.e. flexible intrinsic constraints that can be modified by extrinsic factors during the lifetime of an individual. Note that stereotyped ontogenetic shifts in diet and digestive strategy represent physiological limitations rather than physiological acclimation. Although such changes take place within the lifetime of an organism, they are generally irreversible and do not result from individual adaptations to extrinsic factors during its lifetime. The developmental progressions responsible for ontogenetic changes in diet and digestive strategy vary over evolutionary time.

In general, digestive constraints on diet choice fall into two categories: limitations on the ability to degrade or absorb important components of a potential food (reaction or absorption constraints), and limitations on the ability to process sufficient amounts of a potential food to meet energy or nutrient requirements (processing constraints). Both types can be modified through adaptation to extrinsic factors (see also Chapter 8).

Reaction-absorption constraints

The examples of interactions between diet and digestion given above all fall into the category of reaction or absorption constraints, i.e. limitations on an animal's ability to digest or absorb given food components. The inability of starlings to digest sucrose is an example of a reaction constraint that results from a physiological limitation. It is fixed for the lifetime of an individual. The examples involving copepods and turtles also illustrate reaction constraints, but in these cases the constraints are subject to physiological acclimation and may be modified by extrinsic dietary factors during an individual's lifetime. Induction of intestinal transporters in mice is an absorption rather than a reaction constraint, but, like the reaction constraints in copepods and turtles, it can be modified through physiological acclimation to changes in diet.

Since, in energetic terms, enzymes and intestinal transporters appear to be relatively inexpensive (Kiørboe *et al.* 1986), it is not surprising that reaction-absorption constraints subject to the physiological acclimation appear to be relatively common. It may cost an animal little, both in the energetic sense and the evolutionary sense, to maintain the ability to synthesize broad complements of enzymes and intestinal transporters. Marine suspension-feeding copepods, for example, synthesize amylase despite the relative scarcity of starch in marine phytoplankton (Harris *et al.* 1986). Costs may be further mitigated by the fact that many enzymes and intestinal transporters are inducible, and synthesis and secretion may be increased or decreased in response to the presence or absence of the substrate in the diet.

Digestive processing constraints

While the reaction-absorption constraints discussed above are limitations on an animal's physiological ability to digest or absorb given food components, digestive processing constraints involve limitations on the rates of digestive reactions or absorption, i.e. limitations on the time required for digestive degradation or absorption of given food components. Digestive processing constraints are thus limitations on an animal's throughput rate, defined with respect to its digestive or absorptive rates. Almost by definition they come into play when food quality is limiting, but food availability is not. Limited food availability is an extrinsic constraint on ingestion and processing rate that may supersede physiological, digestive processing constraints.

Constraints on diet selection that result from physiological limitations on digestive processing rate are common. For example, hindgut
fermentation and relatively slow throughput rates allow howler monkeys
(*Alouatta palliata*) to exploit a relatively refractory diet of leaves but
limit their ability to utilize fruits. In contrast, the faster gut throughput
rates of spider monkeys (*Ateles geoffroyi*), a sympatric species, allow
them to specialize on a low-protein, high-energy diet of fruits but limit
their ability to utilize leaves (Milton 1981). Neither species is able to
adopt the diet of the other. Foregut fermenters generally digest fibre
with higher efficiencies than hindgut fermenters, but they do so at the
expense of longer gut throughput times (Parra 1978; Grajal *et al*. 1989)
and by 'sacrificing' the labile components of ingested food to their
microbial symbionts. As a result, large foregut fermenters cannot
process as much food per unit of time as similarly-sized hindgut fermenters and cannot survive on the high-fibre diets that large hindgut
fermenters can utilize (Janis 1976; Demment & Van Soest 1985; Illius
& Gordon, Chapter 8).

Digestive processing constraints can also be modified by physiological
acclimation, although such modifications appear to be the exception
rather than the rule. Diet quality can affect intestinal length in some
birds, most notably the gallinaceous birds in which individuals grow
longer guts in response to seasonally poor diet quality (Leopold 1953;
Moss 1974). Individuals with longer guts can process more food per day
than individuals with shorter guts and digest it with the same efficiency
(Savory & Gentle 1976; Al-Joborae 1980). Increases in gut length with
decreased diet quality have also been observed in deer mice (*Peromyscus
maniculatus*), but the gut changes are relatively small (Green & Millar
1987).

In contrast to reaction-absorption constraints, processing constraints
subject to physiological acclimation appear to be relatively rare. All animals have some ability to alter gut throughput time to some degree in
response to diet. Gut throughput time in red-backed salamanders
(*Plethodon cinereus*), for example, is 1.6 times longer (115 h vs. 70 h) on
a diet of chitinous ants than it is on a diet of less chitinous dipterans
(Jaeger 1990). Variation in throughput time appears to be an especially
important adaptation in animals, like deposit-feeding invertebrates, that
experience wide variation in food quality but comparatively little variation in rate of food encounter or supply (see Chapter 7). It is this ability
that forms the basis for predictions of optimal ingestion and throughput
rates (e.g. Taghon 1981; Dade *et al*. 1990). There are, however, limits

on an animal's ability to increase its throughput rate and still meet its energetic and nutritional requirements.

One way in which physiological limitations on throughput rate may be overcome (to some extent) is by increasing gut volume. An increase in gut volume without a corresponding increase in gut throughput time allows an animal to process more food per unit of time (as in the birds discussed above), but the potential advantages of this tactic for adapting to diet changes may be lessened by the fact that the animal has to support the extra mass of the gut and its contents (Sibly 1981). Increases in gut volume may also take place only at the expense of other structures. Increased gut volume may, for example, limit the proportion of body volume available for reproductive structures (e.g. Forbes 1989) or muscular tissue (e.g. Grajal *et al.* 1989). An animal thus may gain much less, both in the energetic sense and the evolutionary sense, from the ability to vary maximal gut volume or other aspects of gut structure in response to diet changes than from the ability to synthesize broad complements of enzymes and intestinal transporters. That is, doubling the rate of digestion by secreting more enzymes may be far cheaper than doubling the volume of the gut but may return similar rates of absorptive gain.

The fact that digestive processing constraints appear, in general, to be fixed for the lifetime of an individual allows gut morphologies to be substituted for differences in diet in analyses of resource partitioning (e.g. Langer 1986; Bodmer 1991). The linkage between an animal's diet and its digestive strategy is so tight that many characteristics of each often can be inferred easily from gut morphology (e.g. Hume 1982; Penry & Jumars 1990; Bodmer 1991).

MODELLING DIGESTION

Modelling digestion is the mathematical expression of processing constraints. Model development involves explicit treatment of processing patterns and rates (patterns and rates of solid and liquid flow through a gut) in relation to reaction and absorption rates. Consideration of reaction-absorption constraints (i.e. constraints on an animal's abilities to digest or absorb given food components) is secondary in model development but of primary importance in tests of the models. This comparison of two types of digestion models, compartmental models and chemical reactor models, focuses on understanding how they are each structured by processing constraints.

Two general modelling frameworks

Compartmental models or mass-balance box models are widely applied in varied fields that range, for example, from physical oceanography (e.g. Wunsch & Minster 1982) to sediment geochemistry (e.g. Berner 1980) to bioaccumulation of toxicants (e.g. Landrum & Robbins 1990) to digestion. They represent one framework for analyzing complex processes that involves breaking down the process into stages and describing the interactions among the stages (Rosenblatt 1988; Nagle & Saff 1989). The basic unit of these models, a compartment, is characterized by some function $x(t)$ that describes some property x in the compartment at time t, by the rate of flow of x into (I) and out of (E) the compartment and by the rate of production or destruction (R) of x within the compartment. The property x is assumed to be distributed homogeneously throughout the compartment at all times, i.e. the compartment is always well mixed, and the mass balance is written over the entire volume, V, of the compartment (Fig. 3.1).

Chemical reactor models are also mass-balance models (Levenspiel 1972; Penry & Jumars 1987). A general mass-balance equation for x is written over some arbitrary unit of reactor volume: the rate of flow of x into the unit of reactor volume I equals the sum of the rate of outflow (E), the rate of production or destruction (R) and the rate of accumulation (A) of x in the unit of reactor volume. There are three ideal reactor types that differ primarily in the way in which material is processed within the reactor (Fig. 3.1). The batch reactor, as its name implies, processes material in discrete batches. There is no inflow or outflow of material during reaction ($I = E = 0$), and its contents are well mixed. The mass balance is thus written over the entire batch reactor volume V, and changes in x occur only with respect to time.

In contrast to the batch reactor, the continuous-flow, stirred-tank reactor (CSTR) and the plug-flow reactor (PFR) process material continuously ($I, E > 0$). The CSTR is distinguished from the PFR by the fact that its contents are well mixed over the entire reactor; the contents of the PFR are well mixed radially (across the path of material flow), but no mixing occurs longitudinally (parallel to the path of material flow). At steady state the concentration of x is homogeneous throughout the CSTR, and the mass balance is written over the entire CSTR volume V. In contrast there is a gradient in the concentration of x from the inlet to the outlet along the flow path of the PFR, and the mass balance is written over a differential PFR volume element dV. (See Levenspiel 1972 or

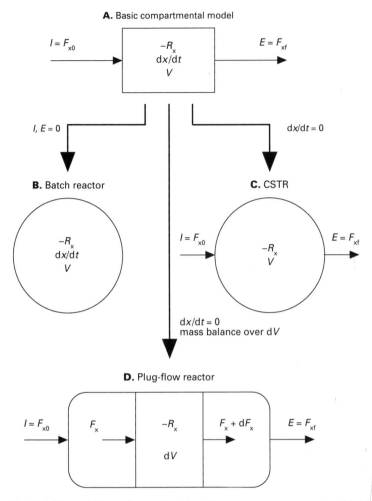

Fig. 3.1 Modelling variables and general relationships among compartmental models (A) and gut-reactor models (B, C and D). *I* and *E* are, respectively, the input and output rates of some component *x* in moles time^{-1}. $-R_x$ is reaction rate of *x* in moles volume^{-1} time^{-1}. A mass balance is written over total volume, *V*, for the compartmental, batch and CSTR models. It is written over a differential volume element, d*V*, for the PFR.

Penry & Jumars 1987 for details of reactor-specific equation derivations.) Plug-flow processing is by far the most common digestive tactic among multicellular animals – most animals have guts that operate at least in part as PFRs (Penry & Jumars 1987; Dade *et al.* 1990). The lack of an explicit compartmental framework for treating plug-flow processing is the greatest, and perhaps the most important, difference between reactor and compartmental modelling approaches.

From these brief descriptions and definitions of compartmental models and reactor models it is clear that compartmental models are not fundamentally different from reactor models. In fact a compartment is either a batch reactor or a CSTR depending on whether I and E are equal to zero (batch processing) or greater than zero (continuous-flow processing), and linkage of ten or more CSTRs or continuous-flow compartments in series closely approximates the behaviour of a PFR (Levenspiel 1972). Thus, as long as the same assumptions and processing constraints are used in model development (e.g. batch vs. continuous flow, mixing vs. no mixing), compartmental models and reactor models can be made interchangeable. In practice, especially in modelling digestion in ruminants, they often are not.

Focus on ruminants

Comparison of compartmental and reactor models will focus on modelling digestion in ruminants, since most compartmental models of digestion have been developed for them (e.g. Brandt & Thacker 1958; Waldo *et al.* 1972; Mertens & Ely 1979; France *et al.* 1990). Analysis of compartmental models for ruminants requires some review (by no means exhaustive) of the history of their development. The models have been elaborated greatly since first proposed at least 30 years ago, but some of the basic modelling constraints and assumptions have remained more or less unchanged to date.

There are two types of compartmental models for ruminants. The first type, the passage model, describes tracer disappearance from the rumen or entire gut to determine mean residence times of particles or fluids (e.g. Blaxter *et al.* 1956; Brandt & Thacker 1958; Ellis *et al.* 1979; Cochran *et al.* 1987). The second type, the digestion model, describes degradation of food materials in and passage from the rumen (e.g. Waldo *et al.* 1972; Mertens & Ely 1979; Aitchison *et al.* 1986; Fisher *et al.* 1989; France *et al.* 1990). The compartmental digestion models incorporate aspects of the compartmental passage models.

Compartmental passage models

Based on observations of the patterns of fecal output of pulses of conservative tracers, the ruminant digestive system is generally modelled as a sequential, irreversible, two-compartment process with continuous flow through the entire system and complete, instantaneous mixing in each compartment (e.g. Ellis *et al.* 1979; Fig. 3.2). In reactor terms this compartmental model is equivalent to two CSTRs in series. The first compartment generally represents the rumen, and the second compartment comprises the remainder of the gut. Elaborations of this basic model to achieve better agreement with observations include the addition of time delays (e.g. Blaxter *et al.* 1956) to account for the fact that

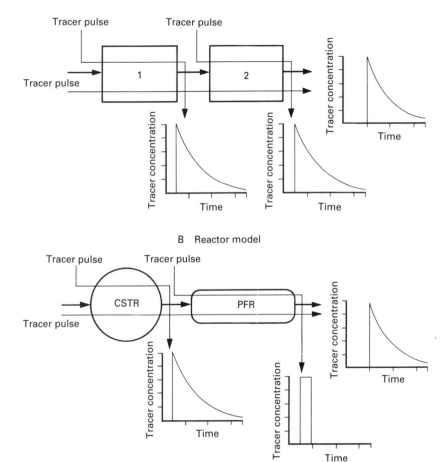

Fig. 3.2 Comparison of tracer residence-time distributions for a compartmental passage model (A) and a gut-reactor model (B) of a ruminant gut. Overall tracer residence-time distributions (rightmost distributions in each panel) are indistinguishable, but within-gut residence-time distributions reveal important differences in particle processing patterns that can result in differing rates of product formation and absorption.

in real animals, unlike ideal models, tracers are not mixed instantaneously throughout the gut. In some second- and third-generation compartmental passage models age-dependent phenomena, e.g. age-dependent particle turnover rates in the first compartment, replace the age-independent passage assumptions of the basic model (e.g. Hughes & Matis 1984).

Compartmental passage models (either basic or with modifications) provide good physical and mathematical descriptions of particle processing in the rumen but, as comparison with a reactor model for the ruminant gut demonstrates, they treat digesta processing incorrectly in the remainder of the gut (Penry & Jumars 1986). The reactor model for the ruminant gut, like the compartmental passage model, is a continuous-flow model with two components (Fig. 3.2). Like the compartmental passage model the first component of the reactor model is a CSTR (the rumen), but unlike the compartmental model the CSTR is followed by a PFR (the remainder of the gut). The CSTR–PFR model provides a better description of, and fit to, observed tracer defecation patterns (Balch 1950) than does the compartmental (= CSTR–CSTR) model (Penry & Jumars 1986).

Does this difference really matter? If the goal is simply empirical description (without biological insight) of overall patterns of tracer defecation by ruminants, then the difference is probably not important. In fact a single compartment or CSTR (perhaps with some time delay factor to account for the lag time between when tracer leaves the rumen and when it is recovered in the feces) might serve just as well as either two-component model. Even for empirical description, however, the CSTR–PFR model is most parsimonious in that it accurately predicts the observations with a minimum of fitted coefficients. If the goal is to model and understand how particles are processed in the ruminant gut to couple the process of digesta passage with the process of digestion (degradation reactions and absorption) then the difference between the compartmental model and the reactor model is very important.

Patterns and rates of mixing and flow affect chemical rates and extents (Levenspiel 1972). For a given overall gut residence time and given reaction kinetics the predicted extents of food degradation can differ significantly between the CSTR–PFR gut and the CSTR–CSTR (= compartmental) gut, with the magnitude and direction of the differences depending on reaction kinetics, and residence time in each gut component (see Figs 5 and 6 in Penry & Jumars 1987; Levenspiel 1972). In general, it is likely that the extent of food degradation and absorption predicted by the CSTR–PFR model will exceed that predicted by the CSTR–CSTR (= compartmental) model, as a consequence of the higher reaction and absorption rates that can be achieved in a PFR than in a CSTR.

One-compartment (single CSTR) passage models can provide reasonable mathematical descriptions of particle residence times in the rumen or in the ruminant gut as a whole, although Hughes and Matis (1984) found that a one-compartment model with age-dependent passage did not fit their data as well as a two-compartment model with age-dependent passage in the first compartment and age-independent passage in the second. Their findings illustrate what is perhaps the major shortcoming of the empirical, compartmental modelling approach. Mathematical modifications of compartmental models for ruminants can provide progressively better fits to data (i.e. with ever-increasing numbers of fitted coefficients or with substitutions of alternative mathematical functions), but it is often difficult to give biological meanings to the mathematical constructs. In contrast, biological processes define the mathematical structure and components of the reactor-theory models. With compartmental modelling the search is often for a biological process that fits a given, empirically-derived mathematical expression, while with reactor modelling it is often for a mathematical expression that describes a given biological process.

The tendency for biology and mathematics to be uncoupled in compartmental models for ruminants should be recognized, but it does not necessarily invalidate the passage models of their applications. Unfortunately, the same cannot be said for many compartmental digestion models for ruminants. As will be seen below, the lack of an explicit biological framework for compartmental digestion models encourages invalid assumptions and incorrect mathematical formulations of many of the digestion models.

Compartmental digestion models

Compartmental models for ruminant digestion are models of processes in the rumen. This basic model structure is a consequence of historical biases, one of the most important of which is an almost exclusive focus on digestive processes occurring in the reticulorumen and neglect of digestive processes in the remainder of the gut. Historically, the primary interest in studies of ruminant digestion and nutrition has been determination of the abilities of ruminants to utilize dietary fibre, and most fibre degradation occurs in the rumen. Ruminant protein nutrition, however, arguably as important as fibre digestion, cannot be studied using a rumen-only model. Protein, in the form of microbial biomass, is 'produced' in the rumen and degraded and absorbed in the small intestine (Ørskov 1982). Thus modelling protein utilization by ruminants requires, at the very least, the CSTR–PFR gut-reactor model.

The goal in this section, however, is not to model protein utilization by ruminants with the CSTR–PFR model. It is instead, to use the general reactor-theory framework to analyse existing compartmental digestion models, thus bowing to the existing bias and limiting discussions to modelling digestion in the rumen.

Some historical background is necessary to understand the development of compartmental digestion models for the rumen. As noted above, most research on ruminants has focused on their abilities to utilize diets of varying fibre contents. Fibre utilization is quantified in terms of fibre digestibility, the fraction of dietary fibre that is degraded in a given amount of time. Two common methods for measuring fibre digestibility are: (1) incubation of rumen fluid (an inoculum of the microbes that mediate fibre degradation) and a known amount of feed in a flask for a set length of time, and (2) placement of a polyester bag containing a known amount of feed directly in the rumen for a set length of time. In both methods digestibility is determined from comparison of the amounts of fibre present before and after incubation. What is even more important about these methods than the digestibility measurements they yield, is the generally unrecognized fact they provide the frame of reference for many compartmental digestion models.

The frame of reference, i.e. where one draws the 'box' for which a mass balance is to be written, differs fundamentally between compartmental digestion models and gut-reactor models. In compartmental models of digestion (as distinct from compartmental models of gut passage or gut evacuation) the box is drawn around a unit mass of ingested food, the frame of reference in the experimental methods used to measure digestibility (Fig. 3.3). Thus a number of compartmental models of digestion in the rumen are in fact compartmental models of digestibility experiments (e.g. Waldo *et al.* 1972; Van Soest 1982; France *et al.* 1990) rather than of digestion in the rumen. One direct consequence of the inappropriate frame of reference used to develop many of the compartmental digestion models is incorrect formulation of the associated sytems of differential equations.

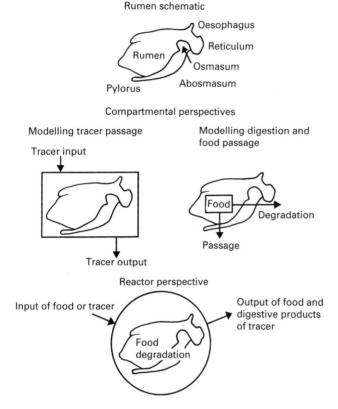

Fig. 3.3 Frames of reference for development of a compartmental passage model, compartmental digestion model, and a CSTR model for the rumen. The modelling 'unit' is the entire rumen in the compartmental passage model and the CSTR model. In contrast the modelling unit is a unit mass of ingested food in the compartmental digestion model.

Compartment volume, a necessary term in compartmental model equations (Fig. 3.1; Rosenblatt 1988; Nagle & Saff 1989), never appears in the equations associated with compartmental digestion models for ruminants.

In gut-reactor models, by contrast, the 'box' is drawn around a unit volume of gut. Since the rumen can be modelled as a CSTR, and the contents of a CSTR are, by definition, perfectly mixed, the box over which the reactor-model mass balance is written is the entire rumen CSTR (Fig. 3.3), and gut reactor volume appears explicitly in model development and equation derivations (Fig. 3.1; Levenspiel 1972; Penry & Jumars 1987).

A second consequence of using a compartmental model of a digestibility experiment to model digestion in the rumen, is the implicit assumption of invalid digesta-processing constraints. The fate of a unit mass of food is followed after it enters the rumen (or digestibility flask or bag). A mass balance is written for the unit mass of food. Food mass can be lost via digestive degradation or via passage out of the gut (Fig. 3.3). If passage is neglected, the problem can be restated in terms of the models illustrated in Fig. 3.1. Food is 'loaded' into the rumen (or digestibility flask or bag), and then digestive degradation (reaction) takes place. Input I equals output E equals zero during degradation – in other words, many 'compartmental' models treat digestive degradation in the rumen as a *batch* process. The implicit assumption of batch processing comes directly from digestibility experiments run in batch mode.

Basic compartmental models are, by definition, continuous-flow systems (Fig. 3.1 and related discussion). The associated sets of differential equations should include, at the very least, terms describing compartmental input and terms describing compartmental output (Rosenblatt 1988; Nagle & Saff 1989). In the case of these rumen models, compartmental input is assumed implicitly and incorrectly to be zero at all times, and input terms are not included in the differential equations (e.g. Waldo *et al.* 1972; Van Soest 1982; France *et al.* 1990). Output terms (i.e. passage terms), however, are included in the models and differential equations (Fig. 3.3). Since inflow is zero, but degradation and outflow can occur, these rumen models are actually semibatch reactors.

Semibatch reactors have some interesting biological applications and implications. Semibatch processing is advantageous in the case where all reactants are added initially to the reactor but one (or more) of the reaction products is (are) removed continuously. This mode of processing results in increased reaction rate due to maintenance of reactant concentrations that are higher than they would be if the reaction products remained in the reactor as reactant diluents (Smith 1981). Animals that process food in batches but absorb digestive products continuously during the processing of each food batch, thereby may be able to compensate for the disadvantages inherent in discontinuous batch processing vs. continuous-flow processing (discussed in Penry & Jumars 1987). This type of semibatch processing, however, is neither assumed nor modelled in the development of compartmental digestion models for the rumen. The outflow term represents outflow of potentially reactive food material, not reaction products.

Comparison of CSTR, batch and semibatch
compartmental models

A very simple exercise demonstrates the differences between a CSTR model of digestion in the rumen, a batch model and a semibatch model. In this exercise it is assumed, as in most models of ruminants, that digestive degradation kinetics in the rumen are first order:

$$-R_x = -k_d \, C_x \qquad (3.1a)$$

in the CSTR and batch models and

$$-R_{-x} = -k_d \, x \qquad (3.1b)$$

in the semibatch compartmental model where the reaction rate coefficient, k_d, has units of time^{-1}, C_x has units of moles volume^{-1} and x has units of moles. The reaction rate equations differ because the necessary volume term is neglected in the development of semibatch compartmental digestion models. It is important to note that digestive degradation reactions in the rumen have autocatalytic kinetics and that, from a biological standpoint, a form of the Michaelis–Menten rate equation, modified for autocatalytic biological reactions (Bischoff 1966), would be preferable to the first-order reaction rate equation. The modified Michaelis–Menten equation, however, makes the mathematics much more complicated and is not undertaken here. The purpose of this exercise is simply to compare three models, not to develop a realistic rumen model.

It is also assumed, as in most rumen models, that passage kinetics are first order:

$$E = -F_{xf} = -k_p \, x \qquad (3.2)$$

where k_p has units of time^{-1} and x units of moles. This expression is a component of the semibatch, compartmental model only. The form of this expression is derived from tracer passage-time studies with ruminants, in which exponential decrease in the relative amount of a conservative tracer in the feces is observed over time after the feeding of a tracer pulse (Fig. 3.2; Balch 1950; Blaxter *et al.* 1956; Ellis *et al.* 1979; Cochran *et al.* 1987). The exponential decrease in relative amount of tracer is characteristic of an animal with a continuous-flow, mixed gut (Penry & Frost 1990). Mixing in the rumen results in continuous dilution of food labelled with a conservative tracer, by subsequently ingested food without tracer.

Equation 3.2 describes relative tracer disappearance from the gut of a continuously feeding ruminant. In compartmental digestion models, however, it is used implicitly to describe the decrease in the mass of a unit of food that results from passage of that mass out of the rumen. There is no evidence that passage of mass out of the rumen when inflow is zero can be described with an exponential decay function. The important point, however, is that an equation derived to describe relative tracer disappearance from a *continuous-flow gut* cannot be plugged uncritically into a model that describes the disappearance of mass from a *batch of food*. Similar confusion of the processes of mixing and dilution within a continuous-flow gut and evacuation of mass from a gut under the condition of no inflow of material, exists in the large body of literature on food processing by suspension-feeding copepod crustaceans (Penry & Frost 1990). As with copepods, mixing and dilution of tracer in the ruminant gut and evacuation of mass from the ruminant gut are two distinct processes. Physiological mechanisms that may relate the two processes or cause them to have similar kinetic descriptions have yet to be documented. Physiological implications aside, the message for developers of compartmental rumen models is clear; the common assumption of semibatch food processing is a poor one.

The differential equation describing the fate of food component x in a semibatch rumen is:

$$dx/dt = -R_{-x} - F_{xf} = -k_d x - k_p x \qquad (3.3a)$$

or

$$t = (-k_d - k_p)^{-1} \ln(x). \qquad (3.3b)$$

The equation associated with the batch reactor model is (after Penry & Jumars 1987):

$$t = x_0 \int_0^{Z_{xf}} dZ_x / -R_x V \qquad (3.4)$$

where t is batch holding or residence time, x_0 is the moles of food component x at time 0, Z_x is the fraction of x degraded and V is reactor volume.

The equation associated with the CSTR model is (after Penry & Jumars 1987):

$$T = C_{x0} Z_x / -R_x \qquad (3.5)$$

where T is reactor throughput or mean residence time and equals the ratio of reactor volume V to volumetric throughput rate, and C_{x0} is the concentration of x in the reactor inflow (moles per unit of volume).

Equations 3.3b, 3.4, and 3.5 can be used to compare the fraction of food component x remaining vs. time under each set of model conditions (Fig. 3.4). It is apparent (Fig. 3.4a) that for any given residence time (with first-order reaction kinetics) the fraction of food degraded will be less in a CSTR than in a batch reactor. This difference is a result of the mixing and dilution and resulting reduction in reactant concentration, and thus in reaction rate, that occurs in CSTR. The fraction of food degraded in a semibatch compartmental model appears to be greater than that degraded in a batch reactor, but the apparently greater extent of degradation is solely a result of the fact that some fraction of undigested food is removed via passage (Fig. 3.4b).

As already stated, compartmental models of digestion in ruminants have generally been developed to compare diets in terms of digestibili-

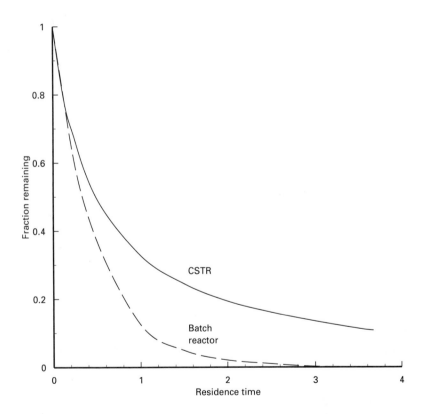

Fig. 3.4a For any given residence time and first-order reaction kinetics (reaction rate coefficient $k_d = 2$ time^{-1}) the fraction of undigested food remaining in a CSTR gut is greater than that remaining in a batch gut.

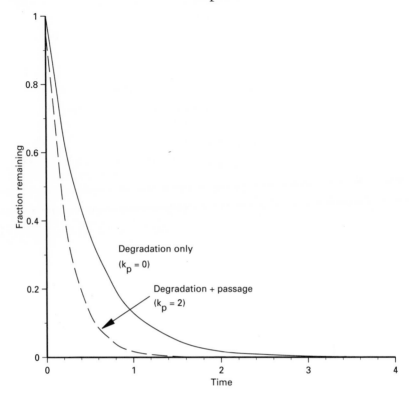

Fig. 3.4b A compartmental digestion model in which input and output are both zero and digestion reactions have first-order kinetics (k_d = 2 time $^{-1}$) is a batch gut. The 'degradation only' curve is identical to the batch reactor curve in (a). A compartmental digestion model in which input is zero, output is nonzero (passage modelled as a first-order process with rate coefficient k_p = 2 time $^{-1}$) and reaction kinetics are first order is a semibatch gut ('degradation + passage'). The apparently greater extent of digestion (i.e. smaller fraction of undigested food remaining at each time) in a semibatch gut is an artifact of the output of undigested food.

ties. The passage constant, k_p, is generally determined from tracer studies (recall that the k_p that is determined from the tracer studies is not the k_p that is called for in the semibatch digestion models, but it is the models that are incorrect, not the tracer studies). The degradation constant, k_d, is generally unknown, and one goal of modelling efforts often is to determine empirically a k_d for a given diet and a given k_p (e.g. Waldo *et al.* 1972; Aitchison *et al.* 1986; Fisher *et al.* 1989). Errors in k_p translate directly into errors in estimates of k_d, but for the sake of this brief analysis it is assumed that k_p can be estimated without error. If the rumen acts as a CSTR (k_{dCSTR} = constant) but is modelled as a

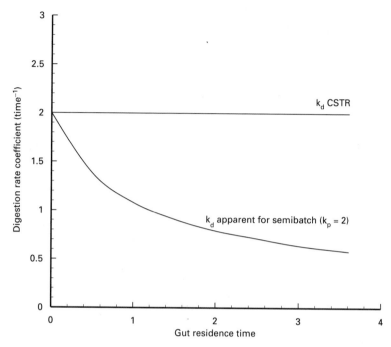

Fig. 3.5 Semibatch compartmental digestion models are often used to estimate k_d, the digestion rate coefficient of a given diet (assuming first-order reaction kinetics), if the passage rate coefficient k_p is known (and first-order passage kinetics are assumed). If a gut operates as a CSTR with first-order reaction kinetics (k_d = constant) but is instead modelled as a semibatch compartment the apparent k_d estimated using the semibatch model declines monotonically with gut residence. The 'apparent k_d' curve was generated by setting the performance equation (Equation 3.5) for a CSTR (with $-R_x = -k_d C_x$ and $k_d = 2$) equal to the performance equation (Equation 3.3b) for a semibatch compartmental model (with $k_p = 2$) and solving for $k_{dapparent}$ for the semibatch model.

batch reactor or semibatch compartment, then the calculated k_{dbatch} or $k_{dsemibatch}$ will decrease monotonically with gut residence time and will always be underestimated with respect to k_{dCSTR} (Fig. 3.5). An estimate of k_{dbatch} or $k_{dsemibatch}$ determined for gut residence time = 1 is about 55% of the value for k_{dCSTR} at gut residence time = 1. Mertens and Ely (1979) suggested (using a continuous-flow compartmental model) that a 1.0% increase in digestion rate can result in a 0.6% increase in maximal digestible dry matter intake. Thus the use of a batch or semibatch compartmental model to estimate k_d and then to predict dry matter intake may underestimate maximal consumption by about one-third as compared to a CSTR model.

MODELLING CONCLUSIONS

Digestive processing patterns and constraints must be stated explicitly at the outset of model development. Most importantly, does the animal process food discontinuously (i.e. does it eat and digest discrete meals) with feeding bouts relatively short compared to gut residence time, or does it ingest and process food more or less continuously with feeding bouts relatively long compared to gut residence time? What are digesta flow patterns within the gut? What is the scale on which digesta mixing takes place? Is there a rather large mixing chamber that should be modelled as a CSTR (e.g. the rumen), or is there a series of relatively small mixing cells that, taken together, can be modelled as a PFR? Identification of the fundamental processing constraints appropriate to a given animal allows one to choose among the ready-made reactor models with their associated performance equations, or to develop a compartmental model and derive correctly the associated equation(s). Failure to identify and state processing constraints is directly responsible for the errors in many compartmental digestion models for ruminants.

From a mathematical standpoint, gut-reactor models and correctly formulated compartmental models can be interchangeable as long as they are structured by the same set of processing constraints. Gut-reactor models, however, have a significant advantage in that they draw upon an extensive body of reactor and chemical process design theory. Compartmental models are generally empirical. The major problem with gut-reactor models, at present, is that theoretical and model development have outpaced abilities to make the measurements necessary to test the predictions (Penry & Jumars 1990; Dade *et al.* 1990). With the models driving methods development, however, this problem will be overcome. The major problem with empirical compartmental models is that their mathematical relationships may be difficult to link explicitly with biological processes. Unlike in gut-reactor models and the explicit iterative procedures developed for them (Penry & Jumars 1987), if important biological parameters are missing in the initial compartmental model formulation, they may never be identified and included.

Mathematical modelling is a useful way to investigate the interplay between diet and digestion. The specific modelling approach is only as good as the conceptual framework for model development. Thus models that only allow prediction of diet digestibilities will always be limited in utility in comparison to models that predict net digestive gains to an animal.

REFERENCES

Aitchison E., Gill M., France J. & Dhanoa M.S. (1986) Comparison of methods to describe the kinetics of digestion and passage of fiber in sheep. *J. Sci. Food Agric.* **37**, 1065–72..

Alexander R. McN. (1991) Optimization of gut structure and diet for higher vertebrate herbivores. *Phil. Trans. R. Soc. Lond. B* **333**, 249–55.

Al-Joborae F.F. (1980) *The influence of diet on the gut morphology of the starling (Sturnus vulgaris L. 1758).* Ph.D. dissertation. University of Oxford, Oxford.

Balch C.C. (1950) Factors affecting the utilization of food by dairy cows. 1. The rate of passage of food through the digestive tract. *Br. J. Nutr.* **4**, 361–88.

Berner R.A. (1980) *Early Diagenesis.* Princeton University Press, Princeton, N.J.

Bischoff K.B. (1966) Optimal continuous fermentation reactor design. *Can. J. Chem. Eng.* **44**, 281–4.

Bjorndal K.A., Suganuma H. & Bolten A.B. (1991) Digestive fermentation in green turtles, *Chelonia mydas,* feeding on algae. *Bull. Mar. Sci.* **48**, 166–71.

Blaxter K.L., Graham N. McC. & Wainman F.W. (1956) Some observations on the digestibility of food by sheep, and on related problems. *Br. J. Nutr.* **10**, 69–91.

Bodmer R.E. (1991) Influence of digestive morphology on resource partitioning in Amazonian ungulates. *Oecologia.* **85**, 361–65.

Brandt C.S. & Thacker E.J. (1958) A concept of rate of food passage through the gastrointestinal tract *J. Anim. Sci.* **17**, 218–23.

Cochran R.C., Adams D.C., Galyean M.L. & Wallace J.D. (1987) Examination of methods for estimating rate of passage in grazing steers. *J. Range Manage.* **40**, 105–8.

Dade W.B., Jumars P.A. & Penry D.L. (1990) Supply-side optimization: maximizing absorptive rates. In *Behavioral Mechanisms of Food Selection* (ed. by R.N. Hughes) *NATO ASI series, vol. G 20,* pp. 531–55. Springer Verlag, Berlin.

Demment M.W. & Van Soest P.J. (1985) A nutritional explanation for body-size patterns of ruminant and nonruminant herbivores. *Am. Nat.* **125**, 641–72.

Ellis W.C., Matis J.H. & Lascano C. (1979) Quantitating ruminal turnover. *Federation Proc.* **39**, 2702–6.

Fisher D.S., Burns J.C. & Pond K.R. (1989) Kinetics of *in vitro* cell-wall disappearance and *in vivo* digestion. *Agron. J.* **81**, 25–33.

Forbes T.L. (1989) The importance of size-dependent processes in the ecology of deposit-feeding benthos. In *Ecology of Marine Deposit Feeders* (ed. by G. Lopez, G. Taghon & J. Levinton), pp. 171–200. Springer Verlag, Berlin.

France J., Thornley J.H.M., Lopez S., Siddons R.C., Dhanoa M.S., Van Soest P.J. & Gill M. (1990) On the two-compartment model for estimating the rate and extent of feed degradation in the rumen. *J. Theor. Biol.* **146**, 269–87.

Grajal A., Strahl S.D., Parra R., Dominguez M.G. & Neher A. (1989) Foregut fermentation in the hoatzin, a neotropical leaf-eating bird. *Science* **245**, 1236–8.

Green D.A. & Millar J.S. (1987) Changes in gut dimensions and capacity of *Peromyscus maniculatus* relative to diet quality and energy needs. *Can. J. Zool.* **65**, 2159–62.

Harris R.P., Samain J.-F., Moal J., Martin-Jézéquel V. & Poulet S.A. (1986) Effects of algal diet on digestive enzyme activity in *Calanus helgolandicus. Mar. Biol.* **90**, 353–61.

Hassett R.P. & Landry M.R. (1990) Effects of diet and starvation on digestive enzyme activity and feeding behaviour of the marine copepod *Calanus pacificus. J. Plankton Res.* **12**, 991–1010.

Hughes T.H. & Matis J.H. (1984) An irreversible two-compartment model with age-dependent turnover rates. *Biometrics* **40**, 501–5.

Hume I.D. (1982) *Digestive Physiology and Nutrition of Marsupials.* Cambridge University Press, Cambridge.

Hume I.D. (1989) Optimal digestive strategies in mammalian herbivores. *Physiol. Zool.* **62**, 1145–63.

Jaeger R.G. (1990) Territorial salamanders evaluate size and chitinous content of arthropod prey. In *Behavioral Mechanisms of Food Selection* (ed. by R.N. Hughes), *NATO ASI series, vol. G 20*, pp. 111–26. Springer Verlag, Berlin.

Janis C. (1976) The evolutionary strategy of the Equidae and the origins of rumen and cecal digestion. *Evolution,* **30**, 757–74.

Jumars P.A. & Penry D.L. (1989) Digestion theory applied to deposit feeding. In *Ecology of Marine Deposit Feeders* (ed. by G. Lopez, G. Taghon & J. Levinton), pp. 114–28. Springer Verlag, Berlin.

Kaiser M.J., Westhead A.P., Hughes R.N. & Gibson R.N. (1992) Are digestive characteristics important contributors to the profitability of prey? A study of diet selection in the fifteen-spined stickleback, *Spinachia spinachia* (L.). *Oecologia,* **90**, 61–69.

Karasov W.H, Pond R.S., III, Solberg D.H. & Diamond J.M. (1983) Regulation of proline and glucose transport in mouse intestine by dietary substrate levels. *Proc. Natl. Acad. Sci. USA* **80**, 7674–7.

Karasov W.H. & Diamond J.M. (1988) Interplay between physiology and ecology in digestion. *BioScience,* **38**, 602–11.

Kiørboe T., Möhlenberg F. & Hamburger K. (1985) Bioenergetics of the planktonic copepod *Acartia tonsa*: relation between feeding, egg production and respiration, and composition of specific dynamic action. *Mar. Ecol. Prog. Ser.* **26**, 85–97.

Landrum P.F. & Robbins J.A. (1990) Bioavailability of sediment-associated contaminants to benthic invertebrates. In *Sediments: Chemistry and Toxicity of In-Place Pollutants* (ed. by R. Baudo, J.P. Giesy & H. Muntau), pp. 237–63. Lewis Publishers, Boston.

Langer P. (1986) Large mammalian herbivores in tropical forests with either hindgut- or forestomach-fermentation. *Z. Säugetierkunde* **51**, 173–87.

Leopold A.S. (1953) Intestinal morphology of gallinaceous birds in relation to food habits. *J. Wildl. Manage.* **17**, 197–203.

Levenspiel O. (1972) *Chemical Reaction Engineering.* John Wiley & Sons, New York.

Martinez del Rio C. & Stevens B.R. (1989) Physiological constraint on feeding behaviour: intestinal membrane disaccharidases of the starling. *Science* **243**, 794–6.

Mertens D.R. & Ely L.O. (1979) A dynamic model of fiber digestion and passage in the ruminant for evaluating forage quality. *J. Anim. Sci.* **49**, 1085–95.

Milton K. (1981) Food choice and digestive strategies of two sympatric primate species. *Am. Nat.* **117**, 496–505.

Mortimer J.A. (1982) Feeding ecology of sea turtles. In *Biology and Conservation of Sea Turtles* (ed. by K. Bjorndal), pp. 103–9. Smithsonian Institution Press, Washington, D.C.

Moss R. (1974) Winter diets, gut lengths, and interspecific competition in Alaskan ptarmigan. *Auk,* **91**, 737–46.

Nagle R.K. & Saff E.B. (1989) *Fundamentals of Differential Equations,* 2nd ed. The Benjamin/Cummings Publishing Company, Redwood City, CA, USA.

Ørskov E.R. (1982) *Protein Nutrition in Ruminants.* Academic Press, London.

Parra R. (1978) Comparison of foregut and hindgut fermentation in herbivores. In *The Ecology of Arboreal Folivores* (ed. by G.G. Montgomery), pp. 205–29. Smithsonian Institution Press, Washington, D.C.

Penry D.L. & Frost B.W. (1990) Re-evaluation of the gut-fullness (gut fluorescence) method for inferring ingestion rates of suspension-feeding copepods. *Limnol. Oceanogr.* **35**, 1207–14.

Penry D.L. & Jumars P.A. (1986) Chemical reactor analysis and optimal digestion. *BioScience* **36**, 310–15.

Penry D.L. & Jumars P.A. (1987) Modeling animal guts as chemical reactors. *Am. Nat.* **129**, 69–96.

Penry D.L. & Jumars P.A. (1990) Gut architecture, digestive constraints and feeding ecology of deposit-feeding and carnivorous polychaetes. *Oecologia* **82**, 1–11.

Rosenblatt J. (1988) A more direct approach to compartmental modeling. *Prog. Food. Nutr. Sci.* **12**, 315–24.

Savory C.J. & Gentle M.J. (1976) Changes in food intake and gut size in Japanese quail in response to manipulation of dietary fibre content. *Br. Poult. Sci.* **17**, 571–80.

Sibly R.M. (1981) Strategies of digestion and defecation. In *Physiological Ecology: An Evolutionary Approach to Resource Use* (ed. by C.R. Townsend & P. Calow), pp. 109–39. Sinauer Associates, Sunderland, MA.

Smith J.M. (1981) *Chemical Engineering Kinetics*, 3rd ed. McGraw-Hill Book Company, New York.

Stephens D.W. & Krebs J.R. (1986) *Foraging Theory*. Princeton University Press, Princeton, N.J.

Taghon G.L. (1981) Beyond selection: optimal ingestion rate as a function of food value. *Am. Nat.* **118**, 202–14.

Van Soest P.J. (1982) The kinetics of digestion. In *Nutritional Ecology of the Ruminant* (ed. by P.J. Van Soest), pp. 211–29. O & B Books, Corvallis, OR.

Waldo D.R., Smith L.W. & Cox E.L. (1972) Model of cellulose disappearance from the rumen. *J. Dairy Sci.* **55**, 125–29.

Wunsch C. & Minster J.-F. (1982) Methods for box models and ocean circulation tracers: mathematical programming and nonlinear inverse theory. *J. Geophys. Res.* **87**, 5647–62.

4: The Psychology of Diet Selection

SARA J. SHETTLEWORTH, PAMELA J. REID and
CATHERINE M.S. PLOWRIGHT

Most issues in behavioural ecology can be approached in terms of three questions. The most fundamental is 'What do animals do?' in the circumstances of interest. Behavioural ecologists often try to understand the answer to this question by posing another, 'What should animals do in these circumstances?' Optimality modelling, illustrated in Chapter 2, seeks to provide answers to this question. Our concern is with a third question, 'How do they do it?' That is to say, what are the mechanisms responsible for behaviour, be it selection of diet, mates, or habitat. Answering this last question is not just a sideline to the central issues in behavioural ecology (Krebs & Davies 1991). Increasingly it seems that modelling efforts need to incorporate features of the species' psychology (or indeed, physiology, *see* Chapter 3) as constraints, as has been done by Kacelnik, Brunner and Gibbon (1990) for time perception and patch leaving. Nor is this enterprise a one-way street, with psychologists merely supplying the information necessary for more realistic models. Optimality models can reveal new questions about animal perception, memory or choice and thereby stimulate new studies of psychological mechanisms (Shettleworth 1987).

The psychology of diet selection is not a subject in itself. Diet selection is a functional category, i.e. one defined in terms of the outcome of behaviour (something is eaten). Psychology has to do with how animals perceive, learn, remember and make choices, i.e. causes or mechanisms for behaviour regardless of its function. Consider a bird attacking an insect. It first has to perceive the insect as an object distinct from its background, and it must possess mechanisms for visual-motor coordination to direct its attack accurately. It has to discriminate prey from inedible or noxious but similar-appearing items and, having attacked, it has to dispatch the prey effectively. Finally, the probability of attack should be influenced not only by the immediate characteristics of the prey but also by the context in which it occurs (Chapters 2 and 9). For example, are better items likely to be nearby? Are there predators or rivals in the neighbourhood? How near is the end of the day? This may require sensitivity to aspects of environmental quality outside the animal's current perception, i.e. memory. Few, if any, of the psychological mechanisms used to solve these problems are unique to diet selection. For example, animals may learn how to handle prey efficiently through reinforcement of the motor patterns that lead most quickly to the reward of ingestion, but the same principles of motor learning might apply to acquiring efficient nest-building or copulatory patterns.

Wanting to provide more than a catalogue of psychological mechanisms and examples of each one's role in diet selection by some species or other, we have concentrated here on a few aspects of diet selection that raise particularly interesting (and in some cases unsolved) psychological problems. Some of our examples, such as search images and social transmission, have been most studied in the context of diet selection. Our theme is that understanding some aspects of diet selection may require very detailed knowledge of psychological mechanisms. Inevitably, our discussion emphasizes common experimental animals – rats, pigeons, fowl, and honeybees – because the extensive studies on them provide the best developed illustrations of this point. We begin with some of the mechanisms involved in assessing the characteristics of individual prey items and then move on to the ways in which the context in which a particular item appears can influence whether it is accepted or rejected.

LEARNING ABOUT PREY

In some species feeding is a reflexive response to specific chemical, visual, or other releasers. For example, frogs strike at stimuli with particular characteristics of size and motion, including 'worms' drawn on cards. The prey-catching response is 'hard-wired' in some amphibians (Sperry 1956). In many species, however, a predisposition to direct feeding responses to certain classes of stimuli is modified by experience, so that some objects within the class of potential prey are accepted while others are ignored or actively avoided (*see also* Chapter 5).

Individual learning: bees

Bees naive to foraging approach flower-like patterns, preferring those of certain shape, colour, and odour. Yet their responses can be modified very quickly if sugar water (artificial nectar) is presented in association with specific colours, shapes, odours or patterns (Gould 1984). Many features of this learning are the same as those found in vertebrates in laboratory studies of Pavlovian (classical) and operant conditioning (Bitterman 1988). But there are two potentially interesting exceptions.

Flowers normally have colour, shape, and odour, and bees normally learn all of these features. When artificial flowers are created lacking odour, it is as if the bee's representation of the flower includes a 'slot' for each sort of information and the odour slot can be filled in later, when a distinctive odour is added (Gould 1984). This is unlike the case with more conventional training paradigms in which initial training with a single cue pre-empts the learning which would otherwise accrue to cues later compounded with it, a phenomenon known as blocking (see Dickinson 1980). Contrary to this claim, however, Bitterman (1988) has reported evidence of blocking in bees.

A second (and equally disputed) aspect of flower learning lies in when the bee learns the association between nectar and the relevant cues (Gould 1984). In most associative learning, the best performance is obtained when a cue precedes the reward. However, bees have sometimes appeared to learn best the cues present as they leave a flower, not those present on the approach. One could argue that it makes sense to process only the features of profitable flowers rather than those of any flower the bee lands on. However, recent experiments indicate that learning can occur during both arrival and departure (Couvillon, Leiato & Bitterman 1991).

Individual learning: fowl

Within a few hours of hatching, young of the domestic fowl or jungle-fowl (*Gallus gallus*) peck at and may ingest a wide variety of small objects, but eventually they peck primarily at food (Hogan 1973). One might think that the discrimination between food and non-food develops like the bees' flower learning, through associating the characteristics of potential food with immediate reward or non-reward. When the immediate effects of ingestion are aversive, this is probably the case (Shettleworth 1972). But the story is more complex when it comes to chicks discriminating foods from non-nutritious objects like sand. For one thing, the discrimination does not appear until the third day after hatching. Moreover, to peck more at food than at sand, three-day-old junglefowl chicks need to have had some experience of pecking followed by ingestion whether or not what was ingested was nutritious (Hogan 1984). This finding can be interpreted as indicating that the motor act of pecking needs to be connected to the chicks' hunger system before the more or less immediate consequences of pecking can modify its rate. The discrimination between food and sand when both are available simultaneously develops gradually and depends on chicks taking a substantial meal of food (Hogan-Warburg & Hogan 1981). Even after learning to identify food, however, chicks still sample a variety of particles.

Social learning: fowl

In natural conditions chicks would feed in the company of a mother hen. Hens have a special food call which attracts the chicks, and while calling they pick up food and drop it in front of the chicks, which may eat it. By drawing the chicks' attention to particular items the hen sets up conditions for the chicks to learn. The hen's 'demonstration' may also tap into another process. If a model hen 'pecks' a spot (pin heads) of a certain colour, chicks separated from her by wire netting selectively peck at spots of that same colour on their side of the netting (Turner 1964). Preference for the colour pecked by the model is retained in the model's absence (Suboski & Bartashunas 1984).

For social factors to have an impact on diet selection, observing conspecifics needs to direct behaviour only until individual learning can take over. For example, in the experiments of McQuoid and Galef (1992) juvenile junglefowl pecked in the same sort of food bowls where they had seen conspecifics feeding. If no food was given in the tests, only a few pecks were directed anywhere, but with food in the tests substantial and enduring preferences for the model's choices developed from these initial socially induced pecking preferences.

Social learning: rats

The discussion of the mechanisms directing food selection in fowl illustrates how the development of diet choice can be canalized by several processes all working together (Hogan 1973). This same principle is illustrated even more vividly by the social transmission of diet choices in rats (Galef 1976; *see also* Chapter 5). A host of different processes conspire to ensure that young rats feed on the same foods as adults in their colony. To begin with, suckling rats acquire a preference for the food flavours transmitted through their mother's milk, and before they leave the nest they also learn about the foods they smell on her fur. Once they leave the nest, young rats prefer to feed where other rats are or recently have been feeding. Through all these routes they are exposed to the flavours and/or odours of foods being safely eaten by others in their colony, and neophobia (fear of unfamiliar foods) ensures that they will continue to choose similar flavours for some time. Exposure to odours of foods other rats have eaten, especially in the context of chemicals from their breath, also transmits information about safe foods among adult rats in a colony (Galef 1990). Of course individual learning, as in taste aversion learning (*see* Chapter 5), can also play a role in diet choice in rats, as in other animals. However, a careful assessment of the evidence indicates that individual learning is inadequate in general to permit selection of a nutritionally adequate diet from a cafeteria of choices (Galef & Beck 1991).

Monitoring profitability

The discussion so far has focussed on some of the ways in which animals come to know whether potential food items are nutritious or poisonous. Deciding what to eat also involves being sensitive to handling time, the delay between encountering an item and ingesting it imposed by the need to process the item in some way. The effects of delay of reinforcement have been much studied in the psychological laboratory and can predict the results of experiments in which animals are confronted with items of different profitabilities (Fantino & Abarca 1985; *see* below). But these are delays imposed by the experimenter, on which the animal's behaviour has no effect. Handling time is under the animal's control to some extent, however, in that it depends on skill in handling the item. This means the profitability of particular items can vary from individual to individual and over time in the same individual as it becomes stronger, larger, or more skilful. In the latter case, unlike the case of information about nutritive value, the animal must continually update its assessment of an item's profitability. There is good evidence that both age and individual or species-specific skills can influence diet choice (Partridge 1976; Laverty & Plowright 1988; Houston, Krebs & Erichsen 1980). As one example, yellow-eyed juncos' relative efficiency at handling large and small mealworms changes with age, and preferences of birds in different age classes can be predicted from age-specific relative profitabilities (Sullivan 1988).

DETECTING AND IDENTIFYING PREY: PERCEPTUAL MECHANISMS

Cryptic prey: experience enhances capture rate

Crypsis is the most common adaptation used by prey to defend against predator detection or identification. By means of colour patterns and structure, cryptic prey can masquerade as a specific item in the environment which is not normally eaten, or can resemble a random sample of background (Endler 1991). The problem faced by the predator is to detect stimuli that deviate in some way from the background pattern and then determine if the stimuli represent something edible. As the time devoted to recognizing a cryptic prey item increases, the costs associated with assessing density and profitability also increase (Erichsen, Krebs & Houston 1980; Hughes 1979). However, with sufficient experience, predators appear to 'break' the crypsis of the prey and improve their rate of detection. L. Tinbergen (1960) was the first to note that, even with no further changes in density, there was often a considerable time lag between the emergence of a cryptic instar and exploitation of the insect as a source of food by songbirds. He proposed that the birds required chance encounters with the new prey before forming a specific 'search image'. This search image was hypothesized to be 'a highly selective sieving operation on the stimuli reaching the retina' (Tinbergen 1960, p. 333). Historically, this sudden specialization on a prey type has been called the search image effect but Hollis (1989) reasonably argues in favour of the more neutral term 'enhanced prey capture'.

Increased probability of capturing a novel cryptic species could imply that the predator is learning any of a variety of things about it (Dawkins 1971a), but predators also show a transitory improvement in the capture of even highly familiar cryptic prey. Pietrewicz and Kamil (1981) used an operant foraging setup to demonstrate this transitory improvement. Well-trained blue jays searched for two cryptic species of *Catocala* moth, each resting on their appropriate bark backgrounds, in slides projected onto a screen. In 'run' sessions of only one species of moth, the birds progressively improved their ability to detect both the presence and the absence of that type. The improvement was transitory because in interleaved 'mixed' sessions of both species, accuracy was poorer and remained stable across trials. Pietrewicz and Kamil suggested that only in run sessions were the birds able to form search images and that concurrent exposure to both species prevented the efficient exploitation of either.

Enhanced prey capture by search image or search rate?

Although Tinbergen's (1960) original conception of search images was rather vague, little empirical research has further illuminated their properties. Indeed, the intuitive appeal of the search image notion seems to have resulted in many behavioural ecologists taking for granted that search images are a proven phenomenon (see Lawrence & Allen 1983 for examples of this). It is generally agreed that search images involve selective attention to certain visual cues (e.g. Dawkins 1971b), or some type of template or mental representation for selective pattern recognition (e.g. Endler 1988). The adaptive value of focussed attention is well documented in research on the processing capacities of humans and other animals (e.g. Johnston & Dark 1986). For example, sticklebacks are less likely to detect an artificial predator when feeding on high-density swarms of waterfleas that are difficult to catch than when feeding on low-density swarms (Milinski 1990). Similarly, starlings are less vigilant for predators when feeding on cryptic baits than when feeding on conspicuous baits, suggesting that cryptic food search is so taxing that vigilance must be foregone (Lawrence 1985). This reasoning, combined with reports of predators specializing on one prey type when two or more are available simultaneously (e.g. Croze 1970; Murton 1971), has led to the widely-held belief that a search image not only improves detection of a specific type but correspondingly reduces the ability of the predator to detect other types of prey that might be present.

Nevertheless, little in the way of convincing evidence exists for improved prey capture by means of search images. Indeed, Gendron and Staddon (1983) showed that simple adjustments in search rate could improve the capture rate of cryptic prey in the absence of search image formation. They argued that the slower a predator searches, the greater the probability of detecting any prey items that are present. On the other hand, the faster a predator searches, the greater the probability of encountering prey items to be detected. As prey become more cryptic, the rate of search must decrease to enhance detection and thereby, maximize capture rate. Observations of pause-travel foragers, such as quail (Gendron 1986) and a variety of planktivorous fish (O'Brien, Browman & Evans 1990), show that pauses lengthen and travel distance decreases when animals are searching for cryptic or hidden prey relative to when searching for more conspicuous prey (Getty & Pulliam, in press).

Search image involves a change in the animal's ability to perceive prey, but search rate adjustment simply involves a change in the rate of scanning an area per unit time. Both mechanisms result in enhanced capture of cryptic prey with repeated experience and both result in specialization on one prey type in multiprey habitat, if one type is encountered more frequently or is easier to detect than the others (Guilford & Dawkins 1987). However, the two hypotheses do differ in what they predict when a predator is faced with equally cryptic prey types. Learning to detect one type of prey by means of a search image is assumed to make the predator less able to detect other cryptic prey types, but adjusting search rate for one type should actually enhance the detection of equally cryptic or more conspicuous types, and such items should be taken in direct proportion to their availability (Guilford & Dawkins 1987).

A new conception of search images

Experiments to distinguish prey detection by search image from prey detection by search rate adjustment were recently reported by Reid and Shettleworth (1992). Pigeons searched for wheat grains dyed brown, green and yellow to mimic different morphs, and these were presented on a substrate of multicoloured gravel. The brown and green grains were equally cryptic and the yellow grains were conspicuous against the same background. When birds fed freely from substrates covered with brown and green grains at various relative densities, they took disproportionately more of the more abundant colour, which suggests that they focussed on the colour most frequently encountered. This is consistent with the search image prediction that the rarer colour would go undetected, but contrary to the search rate prediction that equally cryptic types should be taken in proportion to their availability (Fig. 4.1).

In an operant test resembling those of Pietrewicz and Kamil (1981), Bond and Riley (1991), and Blough (1991), Reid and Shettleworth allowed pigeons to search small plaques of a gravel substrate for single grains which they could eat. When pigeons were switched to a cryptic type in the middle of a session, prior experience with a conspicuous type interfered with detection, but prior experience with a different cryptic type did not (Fig. 4.2, *see* page 65). This result is inconsistent with the prediction that a search image for one type interferes with the detection of other types, but is consistent with adjustment of search rate. In a subsequent experiment, experience with either green or brown grains caused a bias for the matching type in a choice between the two colours when cryptically displayed, but not in a choice between the two colours when

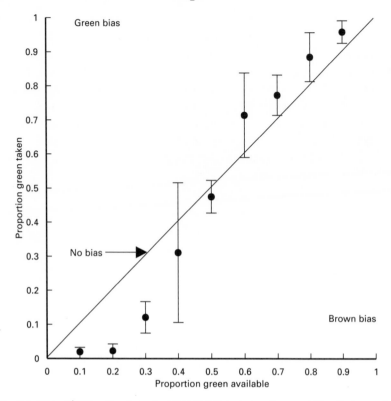

Fig. 4.1 Mean (±SE) proportion of green grains chosen at each proportion of green grains offered in samples of green and brown cryptic grains. The search rate prediction of indifference between green and brown grains is shown with a solid diagonal line. The data, taken from Reid and Shettleworth (in press), are averaged over four birds which each completed 30 trials at each density value. Note that at equal densities, green and brown are chosen equally, as is predicted by both hypotheses if brown and green are equally cryptic.

conspicuously displayed. This result cannot be explained by the search rate hypothesis but is consistent with the search image hypothesis.

These findings suggest that neither the search image nor the search rate hypothesis provides an entirely accurate account of the underlying perceptual mechanism used by predators foraging for cryptic prey. However, the psychological literature on perceptual learning (e.g. McLaren, Kaye & Mackintosh 1989), form perception (e.g. Treisman 1986), and visual search (e.g. Duncan & Humphreys 1989) suggested an explanation of these data in terms of a refined theory of search images. Work on form perception indicates that the visual scene is initially encoded along a number of separable dimensions, such as colour, orientation, brightness, spatial frequency, etc. (Corbetta *et al.* 1990). If the item to be detected is

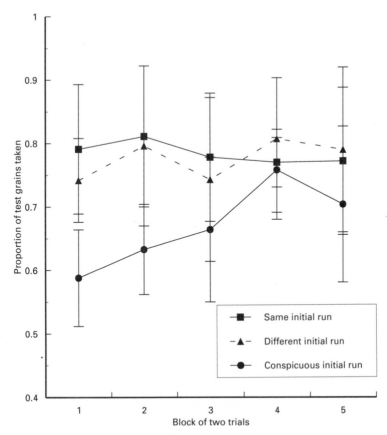

Fig. 4.2 Mean (±SE) proportion of cryptic grains taken following a switch from an initial run (15–17 presentations) of either conspicuous grain, a different cryptic grain, or the same cryptic grain. The data, taken from Reid and Shettleworth (in press), are averaged over five birds which each completed 12 sessions of each initial run type followed by a test run of green grains and 12 sessions of each initial run type followed by a test run of brown grains.

distinctive and differs from all distractors on at least one dimension, such as a green M among yellow Ms, the target 'pops out'. When no single feature can be used to discriminate the target from the distractors, such as a green T among green Ms and blue Ts, attention must be directed toward more than one feature (Treisman & Gelade 1980). As the predator gains experience with the prey against its typical background, it learns which dimensions or features are relevant to the discrimination and becomes increasingly capable of identifying individual items. For example, McLaren, Kaye and Mackintosh (1989) propose that each time a stimulus is experienced, a subset of its features is sampled and associative links are formed between them to comprise the 'central tendency' or representation, of the stimulus. In the case of cryptic prey identification, features which aid in discrimination would be likely to be sampled more frequently than irrelevant features, and thus these distinguishing elements would be more strongly linked. Dawkins (1971b) recognized that selective attention to certain features might be involved in search images and, accordingly, she demonstrated that chicks could discriminate cryptic items on the basis of either colour or shape, whichever was the relevant dimension.

Even when the predator is highly familiar with a cryptic prey type, repeated encounters result in a further transitory improvement in detection by means of sequential priming of the representation (Blough 1991). Priming is thought to be a rapid short-lived process which serves to pre-activate an element in memory, with the result that less activation is required for detection when the actual target is presented (Shiffrin 1988). The stronger the activation, the less attentive processing is required for identification and, in a snowball-like effect, items will continue to be detected more quickly and more accurately.

The advantage of a focus on the analysis of separable features of the prey rather than a more traditional notion of a global 'image' of the prey is that it allows for an explanation of Reid and Shettleworth's data. While activation of a representation for one prey type leads to faster, more accurate detection of that type, the activation can also enhance detection of other types sharing the same features. Repeated experience with one prey type primes all the elements relevant to its discrimination, and if the predator enters a patch containing a different but similar prey type sharing many of the features pertinent to the discrimination, detecting the new type is relatively easy because these shared features are already primed. However, if the types share few features, the predator has difficulty detecting the second type because the relevant elements are not yet activated. Thus, a search image may be more or less specific, depending on the degree of similarity between that type and others present in the environment. While this notion of search images is in line with what is known of both cryptic prey detection and perceptual processes, it needs to be further tested.

Aposematism and mimicry

The interactions of bird predators with brightly coloured and patterned (aposematic) distasteful insects provide one of the most detailed examples of an intimate relationship between the psychology of predators and the evolution of their prey. Whether conspicuousness has evolved as a warning signal to predators, as a thermoregulatory adaptation, or as a result of sexual selection pressures, is a debate that still rages (e.g. Guilford 1989a). However, there is evidence that aposematism does promote quick and long-lasting avoidance learning by predators, either through greater associability of salient (e.g. Gittleman & Harvey 1980) or novel features (Guilford 1989b), or through increasing the initial feeding rates of a naive predator to amplify the unpalatable prey's punishing effects (e.g. Harvey, Bull & Paxton 1981). Aposematism also permits accurate identification from greater distances (Guilford 1989).

Some palatable species fool their predators and avoid being eaten by mimicking the bright colours of distasteful species. Deterministic learning models have been developed to predict the density of Batesian mimics supported by a population of unpalatable models (e.g. Huheey 1988). However, a more psychologically realistic approach involves the application of signal detection theory to foraging for mimetic prey (e.g. Getty, Kamil & Real 1987). Each time a prey-like item is encountered, the predator must decide whether to reject or attack it. If an attacked item is palatable, the predator has scored a 'hit', but if it is unpalatable, the predator has made a 'false alarm'. The frequency of these two outcomes is determined by how choosy the predator is, which is a function of such variables as the relative abundance of, and the degree of similarity between, mimics and models and the aversiveness of models. Research on these topics has been extensively reviewed by others recently, and we refer the interested reader to discussions by Endler (1991), Guilford (1989a), Guilford and Dawkins (1991), and Schuler and Roper (1992).

OPTIMAL DIET SELECTION AND
THE EFFECT OF CONTEXT

The responses of animals to particular events generally depend on the context in which they occur. For example, what patch is chosen depends on whether or not a predator is present (Abrahams & Dill 1989), and how much food starlings collect from a depleting patch depends on the expected rate of energy gain for the whole environment (Cassini, Kacelnik & Segura 1990). Similarly, in diet selection, how an animal treats a particular prey item depends not only on the characteristics of that item and experience with that type of item, but also on contextual variables.

The potential effects of contextual variables on diet selection have been highlighted in models of optimal foraging. For example, whether a forager should accept unprofitable items or reject them in favour of searching further depends on which course of action leads to the higher net rate of energy intake (Stephens & Krebs 1986, but *see also* Chapters 2 and 9).

Two variables which determine which course of action leads to the higher net rate of energy intake are (1) rates of prey encounter (Krebs *et al.* 1977) and (2) time available (Lucas 1990), or 'time horizon' (*see* Chapter 2). In this section we review some evidence that animals are indeed sensitive to overall prey abundance and to time horizon. Also we consider the possible mechanisms underlying this sensitivity. In most of the studies which we describe, operant simulations of diet selection have been used in an effort to discover whether phenomena predicted by foraging models could be obtained in the laboratory and whether they could be explained by principles of operant conditioning. More detailed justification for the use of operant simulations can be found in Lea (1981), Shettleworth (1988), and especially Dallery and Baum (1991).

Sensitivity to rates of prey encounter

A generalized example of an operant simulation of diet selection is shown in Fig. 4.3. A time or response requirement on one key (the 'search key') represents searching for a prey item. Following completion of the search requirement, a second key (the 'prey key') is illuminated with either one of two key colours. One key colour signals a relatively profitable item and the other signals a relatively unprofitable item, where the items may be represented as long vs. short delays to food (Hanson & Green 1989) or long vs. short access to food (Snyderman 1983). The animal can reject an item in favour of searching by responding to the search key or accept it by responding to the prey key.

In operant simulations designed to test the prediction of the effect of changing rate of prey encounter, the search time requirement has been manipulated. According to foraging theory, as search time increases animals should switch from rejecting to accepting all bad items in an all-or-nothing way. An almost universal finding is that animals show partial preferences (but see Snyderman 1983) rather than all-or-nothing acceptance or rejection (for discussion of possible reasons see Shettleworth 1988; McNamara & Houston 1987). Nevertheless, several studies have shown that as search

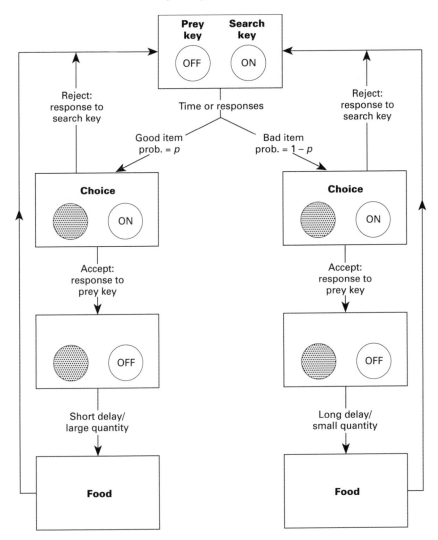

Fig. 4.3 Flow chart of sequence of events in an operant simulation of diet selection.

time increases, the proportion of bad items accepted also increases (Hanson & Green 1989; Fantino & Abarca 1985 for review).

In order to estimate rate of prey encounter, animals must integrate information about the environment over time. A fundamental question in animal learning is what is the time frame (or 'memory window', Cowie 1977) over which this integration takes place. In many studies on foraging, very recent experience influences behaviour disproportionately (e.g. Cuthill *et al.* 1990; Kacelnik & Todd 1992). In diet selection, shore crabs are more likely to accept a relatively unprofitable mussel if they have just encountered and rejected several unprofitable mussels than if they have just consumed a relatively profitable mussel (Elner & Hughes 1978). Shettleworth and Plowright (1992) have recently shown that for pigeons the frequency of acceptance of bad items depended on the very last search time, but did not depend on the search time before last (Fig. 4.4), suggesting that their memory window is very short. Even though the recent past seems to be particularly important in pigeons' responses to prey items, nevertheless the more distant past does enter into their estimate of prey abundance: a higher proportion of bad items was accepted after a given search time when the average search time was 5 sec than when the average was 8 sec (Shettleworth & Plowright unpublished data).

Several models of learning and memory incorporate parameters for the recent and more distant past which are appropriately weighted, albeit often in a post hoc manner (Kacelnik & Krebs 1985). Currently research is being devoted to defining the conditions which promote various weightings of the recent past. For example, Cuthill, Kacelnik and Krebs (in preparation) have suggested that memory window might depend on time horizon: in order to predict events in the immediate future, animals should use the immediate past, and in order to predict events for a long time in the future, animals should integrate over a larger time frame. Unfortunately, as we discuss below, the determinants of animals' sensitivity to time horizon have also only just begun to be explored.

Sensitivity to time horizon

The foraging activity of oystercatchers depends on their tidal regime (Swennen, Leopold & de Bruijn 1989). Foraging shrews are less selective in 3- and 6-min sessions than in 9-min sessions (Barnard & Hurst 1987). Observations such as these suggest that foragers are sensitive to time horizon or some correlate of it. In a test of the prediction that animals should increase their acceptance of bad items toward the end of a foraging bout, Lucas (1987) allowed great tits to encounter meal worms of two different sizes on a conveyer belt. Acceptance increased toward the end of the bouts of 30 sec and 60 sec. In an operant simulation of diet selection with blue jays, Yoerg and Kamil (1988) extended this finding to longer foraging sessions (10 min and 20 min). Pigeons are apparently sensitive not only to session length itself, but to variability in session

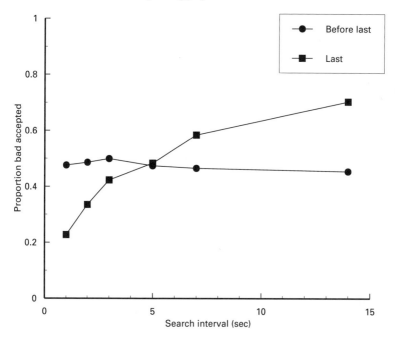

Fig. 4.4 Proportion of bad items accepted as a function of the last interprey interval (search interval) and the one before last by pigeons in an operant diet selection experiment. Adapted from figures for Experiment 1, Block 4 in Shettleworth and Plowright (1992).

length: Plowright and Shettleworth (1991) found that a higher proportion of bad items was accepted in 20-min sessions when the mean session length was 20 min but varied from 5 to 35 min, compared to when the session length was always fixed at 20 min (Fig. 4.5).

The increase in acceptability of bad items at the end of a session may be viewed as an instance of 'anticipatory contrast': the behaviour on a schedule of reinforcement depends on the schedule to follow (Williams & Wixted 1986). In diet selection, perhaps the anticipated end of reinforcement at the end of a session modifies the response to bad items. Whether other models of timing which have been successful in studies of foraging behaviour (as in Kacelnik *et al.* 1990) can also accommodate data on time horizon remains to be seen.

The conditions under which animals demonstrate sensitivity to time horizon have just begun to be investigated (e.g. Lucas, Gawley & Timberlake 1988). Several studies on foraging by pigeons in a patch-choice situation known as the 'two-armed bandit' have failed to show the predicted effect of time horizon on sampling of both alternatives at the beginning of a session (Shettleworth & Plowright 1989). The reasons for this are still

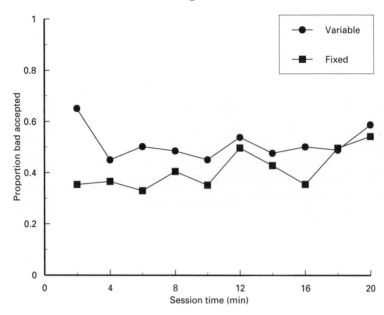

Fig. 4.5 Mean proportion of bad items accepted over 20-min sessions in an operant chamber where sessions were always of fixed 20-min duration and in one where they were of variable duration with a mean of 20 min. From Plowright and Shettleworth (1991), reprinted by permission of the Psychonomic Society, Inc.

unknown. Perhaps Zeiler's (1991) suggestion that within a species, timing of events depends on testing conditions in a way that can be related to the animal's feeding ecology will prove to be a source of insight.

CONCLUSIONS

The apparently simple decision to accept or reject a potential prey item may depend on an enormous amount of acquired information. This information is acquired through a variety of different mechanisms and is relevant on time scales ranging from seconds or minutes – as in the acquisition of a search image – to the animal's lifetime – as in learning about poisonous or unpalatable foods.

With only a few exceptions, our examples have come from studies of birds and mammals. We have tried to show how a detailed knowledge of psychological mechanisms may be necessary for understanding some aspects of diet selection. Those we have mentioned include simple Pavlovian conditioning, priming of visual attention, and sensitivity to time. However, these are only examples. The same diet problems may be solved in different ways by different species. For example, crabs may be struck with assessing prey density by responding to time since the most recent encounter, but other animals may be able to average events over a longer memory window. As we hope we have illustrated, research on diet selection involves a remarkably wide variety of questions about behavioural mechanisms. Many of them will perhaps best be answered by continuing interaction between psychologists and behavioural ecologists (Shettleworth 1984).

REFERENCES

Abrahams M. & Dill L.M. (1989) A determination of the energetic equivalence of the risk of predation. *Ecology* 79, 999–1007.

Barnard C.M. & Hurst J.L. (1987) Time constraints and prey selection in common shrews *Sorex araneus* L. *Anim. Behav.* **35**, 1827–37.

Bitterman M.E. (1988) Vertebrate–invertebrate comparisons. In *Intelligence and Evolutionary Biology* (by H.J. Jerison & I. Jerison), pp. 251–76. Springer Verlag, Berlin.

Blough P.M. (1991) Selective attention and search images in pigeons. *J. Exper. Psychol.: Anim. Behav. Procs.* **17**, 292–8.

Bond A.B. & Riley, D.A. (1991) Searching image in the pigeon: A test of three hypothetical mechanisms. *Ethology* **87**, 203–24.

Cassini M.H., Kacelnik A. & Segura E.T. (1990) The tale of the screaming hairy armadillo, the guinea pig and the marginal value theorem. *Anim. Behav.* **39**, 1030–50.

Corbetta M., Miezin F.M., Dobmeyer S., Shulman G.L. & Peterson S.E. (1990) Attentional modulation of neural processing of shape, colour, and velocity in humans. *Science* **248**, 1556–9.

Couvillon P.A., Leiato T.G. & Bitterman M.E. (1991) Learning by honeybees (*Apis mellifera*) on arrival at and departure from a feeding place. *J. Comp. Psychol.* **105**, 177–84.

Cowie R.J. (1977) Optimal foraging in great tits (*Parus major*). *Nature* **268**, 137–9.

Croze H. (1970) Searching image in carrion Crows. *Zeits. für Tierpsychol.*, Supplement **5**, 1–86.

Cuthill I.C., Kacelnik A., Krebs J.R., Haccou P. & Iwasa Y. (1990) Starlings exploiting patches: The effect of recent experience on foraging decisions. *Anim. Behav.* **40**, 625–40.

Cuthill I.C., Kacelnik A. & Krebs J.R. (in preparation) Looking forward to the past: The Janus effect in starling foraging.

Dallery J. & Baum W.M. (1991) The functional equivalence of operant behaviour and foraging. *Anim. Learning and Behav.* **19**, 146–52.

Dawkins M. (1971a) Perceptual changes in chicks: Another look at the 'search image' concept. *Anim. Behav.* **19**, 566–74.

Dawkins M. (1971b) Shifts of 'attention' in chicks during feeding. *Anim. Behav.* **19**, 575–82.

Dickinson A. (1980) *Contemporary Animal Learning Theory*. Cambridge University Press, Cambridge.

Duncan J. & Humphreys G.W. (1989) Visual search and stimulus similarity. *Psychol. Review* **96**, 433–58.

Elner R.W. & Hughes R.N. (1978) Energy maximization in the diet of the shore crab, *Carcinus maenas. J. Anim. Ecol.* **47**, 103–16.

Endler J.A. (1988) Frequency-dependent predation, crypsis and aposematic coloration. *Phil. Trans. Roy. Soc. Lond.* **319**, 505–23.

Endler J.A. (1991) Interactions between predators and prey. In *Behavioural Ecology: An Evolutionary Approach* (ed. by J.R. Krebs & N.B. Davies), pp. 169–96. Blackwell, London.

Erichsen J.T., Krebs J.R. & Houston A.I. (1980) Optimal foraging and cryptic prey. *J. Anim. Ecol.* **49**, 271–6.

Fantino E. & Abarca E. (1985) Choice, optimal foraging, and the delay-reduction hypothesis. *Behav. and Brain Sci.* 315–30.

Galef B.G., Jr. (1976) Mechanisms for the social transmission of acquired food preferences from adult to weanling rats. In *Learning Mechanisms in Food Selection* (ed. by L.M. Barker, M.R. Best & M. Domjan), pp. 123–48. Baylor University Press, Waco, TX.

Galef B.G., Jr. (1990) An adaptationist perspective on social learning, social feeding, and social foraging in Norway rats. In *Contemporary Issues in Comparative Psychology* (ed. by D.A. Dewsbury), pp. 55–79. Sinauer, Sunderland, MA.

Galef B.G., Jr. & Beck M. (1991) Diet selection and poison avoidance by mammals individually and in social groups. In *Handbook of Behavioral Neurobiology, Vol. 10: Neurobiology of Food & Fluid Intake* (ed. by E.M. Stricker), pp. 329–49. Plenum, New York.

Gendron R.P. (1986) Searching for cryptic prey: Evidence for optimal search rates and the formation of search images in quail. *Anim. Behav.* **34**, 898–912.

Gendron R.P. & Staddon J.E.R. (1983) Searching for cryptic prey: The effect of search rate. *Am. Nat.* **121**, 172–86.

Getty T., Kamil A.C. & Real P.G. (1987) Signal detection theory and foraging for cryptic or mimetic prey. In *Foraging Behaviour* (ed. by A.C. Kamil, J.R. Krebs & H.R. Pulliam), pp. 525–48. Plenum, New York.

Getty T. & Pulliam H.R. (in press) Random prey detection with pause-travel search. *Am. Nat.*

Gittleman J.L. & Harvey P.H. (1980) Why are distasteful prey not cryptic? *Nature* **286**, 149–50.

Gould J.L. (1984) Natural history of honey bee learning. In *The Biology of Learning* (ed. by P. Marler & H.S. Terrace), pp. 149–80. Springer Verlag, Berlin.

Guilford T. (1989a) The evolution of aposematism. In *Insect Defenses: Adaptive Mechanisms and Strategies of Prey and Predators* (ed. by D.L. Evans & J.O. Schmidt), pp. 23–61. New York State University of New York Press, New York.

Guilford T. (1989b) Studying warning signals in the laboratory. In *Ethoexperimental Approaches to the Study of Behavior* (ed. by R.J. Blanchard *et al.*), pp. 87–103. Kluwer Academic Press, Dordrecht & Boston.

Guilford T. & Dawkins M.S. (1987) Search images not proven: A reappraisal of recent evidence. *Anim. Behav.* **35**, 1838–45.

Guilford T. & Dawkins M.S. (1991) Receiver psychology and the evolution of animal signals. *Anim. Behav.* **42**, 1–14.

Hanson J. & Green L. (1989) Foraging decisions: prey choice by pigeons. *Anim. Behav.* **37**, 429–43.

Harvey P.H., Bull J.J. & Paxton R.J. (1981) Looks pretty nasty. *New Scientist* Nov./Dec.

Hogan J.A. (1973) How young chicks learn to recognize food. In *Constraints on Learning* (ed. by R.A. Hinde & J. Stevenson-Hinde), pp. 119–39. Academic Press, London.

Hogan J.A. (1984) Pecking and feeding in chicks. *Learning and Motiv.* **15**, 360–76.

Hogan-Warburg A.J. & Hogan J.A. (1981) Feeding strategies in the development of food recognition in young chicks. *Anim. Behav.* **29**, 143–54.

Hollis K.L. (1989) In search of a hypothetical construct: A reply to Guilford and Dawkins. *Anim. Behav.* **37**, 162–63.

Houston A.I., Krebs J.R. & Erichsen J.T. (1980) Optimal prey choice and discrimination time in the great tit (*Parus major* L.). *Behav. Ecol. and Sociobiol.* **6**, 169–75.

Hughes R.N. (1979) Optimal diets under the energy maximization premises: The effects of recognition time and learning. *Am. Nat.* **113**, 209–21.

Huheey J.E. (1988) Mathematical models of mimicry. In *Mimicry and the Evolutionary Process* (ed. by L.P. Brower), pp. 22–41. University of Chicago Press, Chicago.

Johnston W.A. & Dark V.J. (1986) Selective attention. *Ann. Rev. Psychol.* **37**, 43–75.

Kacelnik A., Brunner D. & Gibbon J. (1990) Timing mechanisms in optimal foraging: Some applications of scalar expectancy theory. In *Behavioural Mechanisms of Food Selection* (ed. by R.N. Hughes), *NATO ASI series, vol. G 20*, pp. 61–82. Springer Verlag, Berlin.

Kacelnik A. & Krebs J.R. (1985) Learning to exploit patchily distributed food. In *Behavioral Ecology* (ed. by R.M. Sibley & R. Smith), pp. 189–205. British Ecological Society, 25th Symposium. Blackwell Scientific Publications, Oxford.

Kacelnik A. & Todd I.A. (1992) Psychological mechanisms and the marginal value theorem; effect of variability in travel time on patch exploitation. *Anim. Behav.* **43**, 313–22.

Krebs J.R. & Davies N.B. (1991) Introduction. In *Behavioral Ecology* (3rd Edition) (ed. by J.R. Krebs & N.B. Davies), pp. ix–x. Blackwell Scientific, Oxford.

Krebs J.R., Erichsen J.T., Webber M.I. & Charnov E.L. (1977) Optimal prey selection in the great tit (*Parus major*). *Anim. Behav.* **25**, 30–38.

Laverty T.M. & Plowright R.C. (1988) Flower handling by bumblebees: A comparison of specialists and generalists. *Anim. Behav.* **36**, 733–40.

Lawrence E.S. (1985) Vigilance during 'easy' and 'difficult' foraging tasks. *Anim. Behav.* **37**, 157–60.

Lawrence E.S. & Allen J.A. (1983) On the term 'search image'. *Oikos* **40**, 313–14.

Lea S.E.G. (1981) Correlation and contiguity in foraging behaviour. In *Predictability, Correlation and Contiguity* (ed. by P. Harzem & M.D. Zeiler), pp. 355–405. John Wiley & Sons, New York.

Lucas G.A., Gawley D.J. & Timberlake W. (1988) Anticipatory contrast as a measure of time horizons in the rat: Some methodological determinants. *Anim. Learn. and Behav.* **16**, 377–82.

Lucas J.R. (1987) The influence of time constraints on diet choice of the great tit, *Parus major. Anim. Behav.* **35**, 1538–48.

Lucas J.R. (1990) Time scale and diet choice decisions. In *Behavioural Mechanisms of Food Selection* (ed. by R.N. Hughes), *NATO ASI Series, vol. G 20*, pp. 165–86. Springer Verlag, Berlin.

McLaren I.P.L., Kaye H. & Mackintosh N.J. (1989) An associative theory of the representation of stimuli: applications to perceptual learning and latent inhibition. In *Parallel Distributed Processing: Implications for Psychology and Neurobiology* (ed. by R.G. Morris), pp. 102–30. Clarendon Press, Oxford.

McNamara J.M. & Houston A.I. (1987) Partial preferences and foraging. *Anim. Behav.* **35**, 1084–99.

McQuoid L. & Galef B.G., Jr. (1992) Social influences on feeding site selection by Burmese fowl (*Gallus gallus*). *J. Comp. Psychol.* **106**, 137–41.

Milinski M. (1990) Information overload and food selection. In *Behavioural Mechanisms of Food Selection* (ed. by R.N. Hughes), *NATO ASI series, vol. G 20*, pp. 721–36. Springer Verlag, Berlin.

Murton R.K. (1971) The significance of a specific search image in the feeding behavior of the wood-pigeon. *Behaviour* **40**, 10–42.

O'Brien W.J., Browman H.I. & Evans B.I. (1990) Search strategies of foraging animals. *Amer. Scientist* **78**, 152–60.

Partridge L. (1976) Individual differences in feeding efficiencies and feeding preferences of captive great tits. *Anim. Behav.* **24**, 230–40.

Pietrewicz A.T. & Kamil A.C. (1981) Search images and the detection of cryptic prey: An operant approach. In *Foraging Behavior: Ecological, Ethological, and Psychological Approaches* (ed. by A.C. Kamil & T.D. Sargent), pp. 311–31. Garland, New York.

Plowright C.M.S. & Shettleworth S.J. (1991) Time horizon and choice by pigeons in a prey-selection task. *Anim. Learning and Behav.* **19**, 103–12.

Reid P.J. & Shettleworth S.J. (1992) Detection of cryptic prey: Search image or search rate? *J. Exper. Psychol.: Anim. Behav. Procs.* **18**, 273–86.

Schuler W. & Roper T.J. (1992) Responses to warning coloration in avian predators. *Adv. Stud. Behav.* **21**, 111–46.

Shettleworth S.J. (1972) The role of novelty in learned avoidance of unpalatable prey by domestic chicks. *Anim. Behav.* **20**, 29–35.

Shettleworth S.J. (1984) Learning and behavioural ecology. In *Behavioural Ecology* (2nd Edition) (ed. by J.R. Krebs & N.B. Davies), pp. 170–94. Blackwell Scientific, Oxford.

Shettleworth S.J. (1987) Learning and foraging in pigeons: effects of handling time and changing food availability on patch choice. In *Quantitative Analysis of Behavior, Vol. 6: Foraging* (ed. by M.L. Commons, A. Kacelnik & S.J. Shettleworth), pp. 115–32. Erlbaum, Hillsdale, N.J.

Shettleworth S.J. (1988) Foraging as operant behavior and operant behavior as foraging: What have we learned? In *The Psychology of Learning and Motivation: Advances in Research and Theory, Vol. 22* (ed. by G. Bower), pp. 149. Academic Press.

Shettleworth S.J. & Plowright C.M.S. (1989) Time horizons of pigeons on a two-armed bandit. *Anim. Behav.* **37**, 610–23.

Shettleworth S.J. & Plowright C.M.S. (1992) How pigeons estimate rates of prey encounter. *J. Exper. Psychol.: Anim. Behav. Procs.* **18**, 219–35.

Shiffrin R.M. (1988) Attention. In: *Stevens' Handbook of Experimental Psychology, Vol. 2* (ed. by R.C. Atkinson, R.J. Herrnstein, G. Lindzey & R.D. Luce), pp. 739–811. John Wiley & Sons, New York.

Snyderman M. (1983) Optimal prey selection; partial selection, delay of reinforcement and self control. *Behav. Anal. Lett.* **3**, 131–47.

Sperry R.W. (1956) The eye and the brain. *Sci. Amer.* **194**, 48–52.

Stephens D.W. & Krebs J.R. (1986) *Foraging Theory.* Princeton University Press, Princeton, New Jersey.

Suboski M.D. & Bartashunas C. (1984) Mechanisms for social transmission of pecking preferences to neonatal chicks. *J. Exper. Psychol.: Anim. Behav. Procs.* **10**, 182–94.

Sullivan K.A. (1988) Age-specific profitability and prey choice. *Anim. Behav.* **36**, 613–15.

Swennen C., Leopold M.F. & de Bruijn L.L.M. (1989) Time-stressed oystercatchers, *Haematopus ostralegus*, can increase their intake rate. *Anim. Behav.* **38**, 8–22.

Tinbergen L. (1960) The natural control of insects in pinewoods. I. Factors influencing the intensity of predation by song birds. *Arch. Néerl. Zool.* **13**, 265–343.

Treisman A.M. (1986) Features and objects in visual processing. *Sci. Amer.* **255**, 106–15.

Treisman A.M. & Gelade G. (1980) A feature-integration theory of attention. *Cog. Psychol.* **12**, 97–136.

Turner E.R.A. (1964) Social feeding in birds. *Behaviour* **24**, 1–46.

Williams B.A. & Wixted J.T. (1986) An equation for behavioral contrast. *J. Exper. Anal. Behav.* **45**, 47–62.

Yoerg S.I. & Kamil A. (1988) Diet choice of blue jays (*Cyanocitta cristata*) as a function of time spent foraging. *J. Comp. Psychol.* **102**, 230–5.

Zeiler M.D. (1991) Ecological influence on timing. *J. Exper. Psychol.: Anim. Behav. Procs.* **17**, 13–25.

5: Foraging as a Self-Organizational Learning Process: Accepting Adaptability at the Expense of Predictability

FREDERICK D. PROVENZA and
RICHARD P. CINCOTTA

INTRODUCTION

The world is constantly changing, yet people often regard change as anomalous, a transitory disruption in a normally predictable world. But change is the rule, not an exception to the rule. Change is the basis upon which the present has emerged from the past, and upon which the future will emerge from the present.

Adaptation results from variation and selection in a changing environment. From this standpoint, there are two ways to explain foraging behaviour. Causal or mechanistic explanations for behaviour (e.g. learning) define the mechanisms by which behaviours are generated (rules of variation) and the mechanisms through which adaptive variants are selected (rules of selection) (Staddon 1983). In contrast, functional explanations for behaviour (e.g. optimality theory) define selection rules only in terms of final outcomes (Staddon 1983). Thus, functional models of foraging are concerned with behaviours that represent goals (Chapter 2), but they do not consider the adaptive processes involved in attaining these goals.

Functional models may only be used to define optimal solutions to particular problems, but they do not, for instance, explain why: (i) individuals within species select different kinds and amounts of forages (Provenza & Balph 1988, 1990), (ii) wild and domesticated herbivores over-ingest plants that contain toxins (Provenza *et al.* 1992), or (iii) herbivores do not necessarily select foods of the highest nutritional quality (e.g. the most energy-rich foods) when given a choice (Grovum 1988). These empirical observations are not aberrations, and are paradoxical only from the perspective of rather rigid explanations that ignore feedback mechanisms (learning) in diet selection.

In this chapter we: (i) review what is known about the role of feedback in foraging behaviour, (ii) discuss some reasons why it has been difficult for functional models of foraging to explain variation and change, and (iii) present a relatively simple simulation model of how learning affects foraging.

ROLES OF FEEDBACK IN FORAGING BEHAVIOUR

Learning provides the feedback necessary for animals to adapt quickly to an ever-changing environment. There are two kinds of feedback mechanisms that have been shown experimentally to be important in foraging: (i) learning from post-ingestive feedback, and (ii) learning from conspecifics, which involves transgenerational interactions.

Post-ingestive feedback

Both wild and domesticated herbivores select nutritious diets from rangelands that contain a diverse array of plant species, individuals, growth stages, and parts that vary in nutritional value, chemical, and mechanical defences (Provenza & Balph 1990). These animals select nutritious diets even though their requirements vary with age, physiological state and environmental conditions (Chapter 8). The ability to select a diet does not occur by chance alone, but is the result, in part, of post-ingestive feedback from nutrients and toxins. Thus, 'palatability' is at least partially a function of the 'nutrient content' of foods, and 'unpalatability' is at least in part related to the 'toxin' concentration of foods. As discussed later in this chapter, experiences early in life also influence what is and is not 'palatable' to herbivores.

Animals increase intake of foods paired with nutrients, presumably on the basis of positive post-ingestive feedback. For example, sheep strongly preferred flavours that were paired with glucose to flavours that were paired with the non-nutritive sweetener saccharin (Burritt & Provenza 1992). This was probably the result of positive post-ingestive feedback because lambs initially exhibited equal preference for the glucose and saccharin solutions, and they did not develop an aversion to the saccharin solution during conditioning. Cattle also apparently develop a preference for supplemental protein blocks when ingesting forage low in protein (Provenza *et al.* 1983). Rats, the mammalian species studied most in this regard, increased their intake of foods or non-nutritive flavours that were paired with: (i) calories (e.g. Booth 1985; Mehiel and Bolles 1984, 1988; Gibson and Booth 1989), (ii) recovery from nutritional deficiencies (e.g. Garcia *et al.* 1967; Zahorik *et al.* 1974; Baker *et al.* 1987; Baker

& Booth 1989), and (iii) recovery from post-ingestive distress (Green & Garcia 1971).

Herbivores decrease intake of foods that contain toxins, presumably on the basis of aversive post-ingestive feedback. As toxicity increases (e.g. aversive feedback from a toxin or from an excess of energy or nutrients), intake of the food decreases. Conversely, as toxicity diminishes and nutritional value improves (i.e. positive feedback from an energy- or nutrient-rich food), intake of the food increases (sheep: Thorhallsdottir *et al.* 1987; Burritt & Provenza 1989a, 1992; du Toit *et al.* 1991; Launchbaugh & Provenza 1993; goats: Provenza *et al.* 1993; Distel & Provenza 1991; cattle: Pfister *et al.* 1990).

Animals also decrease intake of foods deficient in essential nutrients, which are in essence slow-acting toxins, and increase intake of foods that rectify deficiencies (see Rozin & Kalat (1971) and Rozin (1976) for a detailed discussion of this subject). Lambs reduced intake of foods paired with deficits of essential amino acids, and increased intake of foods paired with recovery from such deficits (Rogers & Egan 1975; Egan & Rogers 1978). These data may help to explain an unusual observation. Provenza (1977) studied the nutritional responses of Angora goats to blackbrush (*Coleogyne ramosissima*), a shrub low in nitrogen (0.67%). During the 2-month study, the goats consumed an enormous woodrat (*Neotoma lepida*) house, made of dried juniper (*Juniperus osteosperma*) bark and twigs covered with urine (nitrogen), perhaps because the additional nitrogen provided positive post-ingestive feedback.

The fact that animals can associate the flavour of foods with post-ingestive feedback raises an important, but not fully resolved, question: How do they learn which foods cause which consequences? In the case of conditioned food aversions in rats and ruminants, several variables are involved including: (i) food novelty (Kalat & Rozin 1973; Provenza *et al.* 1990; Burritt & Provenza 1991), (ii) the intensity of the taste as measured by concentration of a flavour (Dragoin 1971; Launchbaugh *et al.* 1993), (iii) the relative amounts of two foods ingested (Bond & DiGiusto 1975; Provenza *et al.* 1993, *see* below), (iv) the temporal sequence of food ingestion (Domjan & Bowman 1974; Bond & DiGiusto 1975), (v) prior experience with illness (Cannon *et al.* 1975), and (vi) prior experience with a salient flavour (Launchbaugh & Provenza 1993; Provenza *et al.* 1993).

The foregoing evidence that herbivores learn from post-ingestive feedback assumes implicitly that they remember what they have learned.

If not, at every moment in time herbivores would be encountering the foraging environment for the first time. It is clear that herbivores: (i) remember specific foods that provided either aversive (Burritt & Provenza 1989b, 1990; Lane *et al.* 1990; Distel & Provenza 1991) or positive (Green *et al.* 1984; Squibb *et al.* 1990; Distel & Provenza 1991) consequences for at least 1–3 years, and (ii) discriminate between novel and familiar foods and sample novel foods cautiously (Chapple *et al.* 1987a,b; Thorhallsdottir *et al.* 1987, 1990a,b; Burritt & Provenza 1989a, 1991; Provenza *et al.* 1990). Sheep and goats decrease intake of novel foods in conditioned food aversion experiments (Burritt & Provenza 1989a; Provenza *et al.* 1990), given as many as four familiar foods and one novel food and a delay between ingestion and consequences of up to 6–8 hours (Burritt & Provenza 1991).

It is important that herbivores sample foods, and that they do not omit completely from their diets plants that contain toxins, because plants vary in nutritional value and toxicity over days, weeks and months. As a result of sampling, herbivores: (i) ingest a diverse array of foods, which reduces their likelihood of over-ingesting a toxin and enhances the likelihood of their meeting all of their nutrient requirements, (ii) can change intake of plants that contain toxins as plant toxicity changes, and (iii) discover new forages.

Subtle changes in flavour, for instance caused by a change in plant chemistry, cause familiar foods to be sampled cautiously (Thorhallsdottir *et al.* 1987; Burritt & Provenza 1989a; Launchbaugh & Provenza 1993). If changes in flavour are associated with a toxin, intake of the food decreases. For instance, lambs made ill after ingesting six familiar foods, subsequently decreased intake of the one food whose flavour was changed (Provenza unpublished).

When animals encounter only novel foods, various factors may enable them to learn which foods are nutritious and which foods are toxic. To understand better the mechanisms that ruminants use to distinguish between novel foods that differ in postingestive consequences, we offered goats current season's (CSG) and older (OG) growth twigs from the shrub blackbrush. CSG is higher than OG in nitrogen (1.04% *versus* 0.74%) and it is more digestible *in vitro* in goat rumen fluid (48% *versus* 38%). Nevertheless, goats acquire a strong preference for OG over CSG because CSG is much higher than OG in a condensed tannin that causes a learned food aversion in goats (Provenza *et al.* 1990).

When CSG and OG were offered to goats naive to blackbrush, the goats did not choose either OG or CSG exclusively, but when they finally ate enough CSG within a meal to experience malaise, they ingested less CSG than OG from that point onward. The change in diet selection resulting from postingestive feedback occurred within hours, not days or weeks, and it was influenced by the "volume of food ingested" during a meal. The greater the volume of a particular food that was ingested, the stronger was the acquired aversion. "Salience" also influenced the response of the goats, and caused a stronger aversion to CSG than to OG. Salience was evidently caused by a flavour common to both OG and CSG, but more concentrated in CSG than in OG.

Insights into acquired preferences and aversions can be gained by careful analysis of diet selection within meals. Such analyses, combined with an understanding of the role of ontogeny in diet selection (discussed below), may simplify understanding of what appears to be a complex process: how ruminants determine which foods are nutritious and which foods are toxic given a diverse array of plant species, individuals, growth stages, and parts that vary in nutritional value and chemical defenses (Provenza and Balph 1990).

Relationship between the senses and post-ingestive feedback

Feedback from eating foods involves affective and cognitive processes (Garcia 1989). Taste plays a prominent role in both processes. Affective processes integrate the **taste** of food and its **post-ingestive consequences**, and changes in the intake of food items depend on whether the post-ingestive consequences are aversive or positive. The net result is called **incentive modification**. On the other hand, cognitive processes integrate the **odour** and **sight** of food with its **taste**. Animals differentiate among foods by smell and sight, and select or avoid foods according to their post-ingestive consequences. The net result is a **change in their behaviour**. The anatomical and physiological mechanisms underlying affective and cognitive systems are discussed for herbivores by Provenza *et al.* (1992).

The senses of taste and smell play important, but somewhat different, roles in diet selection (Garcia 1989). The different roles of odour and taste are illustrated by the fact that a novel odour must be followed immediately by aversive feedback to produce strong odour-aversion learning, but strong aversions to novel tastes can be conditioned even when aversive feedback is delayed up to 12 hours (Rozin 1976; Zahorik & Houpt 1991; Burritt & Provenza 1991). When odour is combined with a distinctive taste, however, the conditioning associated with the odour is enhanced markedly. This effect, called **potentiation**, reflects the fact that a previously weak odour cue becomes a strong associative cue when it is presented in conjunction with taste.

Affective and cognitive processes are influenced by intraspecific variation in anatomy and physiology. Williams (1978) was convinced that individuals are distinctly different in every particular, and that this was the basis of individuality in people:

> Stomachs, for example, vary in size, shape and contour ... They also vary in operation.... A Mayo Foundation study of about 5000 people who had no known stomach ailment showed that the gastric juices varied at least a thousand fold in pepsin content. The hydrochloric-acid content varies similarly.... Such differences are partly responsible for the fact that we tend not to eat with equal frequency or in equal amounts, nor to choose the same foods...

Like people, herbivores apparently vary in their intake of specific foods as a result of intraspecific variation in responses to toxins, and probably nutrients as well. For instance, some sheep fed a high (2.5×) level of *Galega officinalis* failed to show any clinical symptoms of toxicosis, while others were killed by a low (1.0×) dose (Keeler *et al.* 1988). Sheep showed similar variation in susceptibility to toxins in *Verbesina encelioides* (Keeler *et al.* 1992), and goats varied in their ability to ingest condensed tannins in blackbrush (Provenza *et al.* 1990). Different responses are likely, due in part to differences in concentration of the many enzymes required for detoxification and digestion.

Learning from social models

Learning results in foraging information being passed from experienced to inexperienced foragers, for example, from a mother to her offspring. There is ample evidence that mother markedly affects the establishment (e.g. Lobato *et al.* 1980; Key & MacIver 1980; Lynch *et al.* 1983; Thorhallsdottir *et al.* 1990b; Mirza & Provenza 1990, 1992) and the persistence (e.g. Green *et al.* 1984; Lynch 1987; Thorhallsdottir *et al.* 1990a; Nolte *et al.* 1990) of her offspring's dietary habits.

The effects of a mother on her offspring's dietary habits apparently begin *in utero* and continue long after weaning. Rats form preferences (Hepper 1988) and aversions (Stickrod *et al.* 1982; Smotherman 1982) to food flavours based on experience *in utero*. There are even sensitive periods for learning *in utero* (Hill & Przekop 1988). The fetal taste system of lambs is functional during the last trimester of gestation (Bradley & Mistretta 1973), and fetal taste experiences may affect adult food preferences in sheep (Bradley & Mistretta 1973; Hill & Mistretta 1990). Foods ingested by the mother also influence the chemical composition and flavour of her milk (Bassette *et al.* 1986; Panter & James 1990), which subsequently affects intake of solid food in rats (Galef & Sherry 1973; Capretta & Rawls 1974) and ruminants (Morrill & Dayton 1978; Nolte & Provenza 1991). Thus, the food preferences of young livestock are being conditioned before they ever begin to eat solid food.

As young animals make the transition from suckling to foraging, they learn which foods to eat and which to avoid. Learning efficiency increases when a young herbivore's mother is involved. In the absence of the mother, individual lambs offered two novel foods, one safe and the other harmful, did not detect the harmful food (Burritt & Provenza 1989a). By contrast, lambs learned quickly to avoid a harmful novel food, and to select a nutritious novel alternative, when they were with their mothers who exhibited the appropriate behaviour (Mirza & Provenza 1992). Young rats did not select a nutritious diet from an assortment of nutritious and toxic foods when experienced social models were absent (Galef 1985a; Galef & Beck 1990).

As a young animal ages, it depends less on its mother for milk, and a mother apparently has less direct influence on her offspring's dietary habits (Hinch *et al.* 1987; Mirza & Provenza 1990, 1992). At that time, young companions markedly influence one another's intake of food. For instance, lambs who had been averted to the forage shrub *Cercocarpus montanus* (by pairing its ingestion with the toxin lithium chloride administered orally in capsules) consumed more *C. montanus* when they foraged with non-averted lambs than when they foraged alone (Provenza & Burritt 1991). Social influences also ameliorated aversions in lambs fed a pelleted diet in pens (Thorhallsdottir *et al.* 1990c), in heifers grazing on pastures (Lane *et al.* 1990; Ralphs & Olsen 1990), and in rats (Galef 1985b, 1986), which suggests that social influences are a major variable controlling diet selection.

Transgenerational interactions

Socially mediated behaviours cause foraging traditions in animals. Both wild (e.g. Festa-Bianchet 1986; Cederlund *et al.* 1987; Gasaway *et al.*

1989) and domesticated (e.g. Hunter & Milner 1963; Key & MacIver 1980; Zimmerman 1980; Roath & Krueger 1982; Lynch 1987) ruminants form attachments to specific forages and environments that are transferred from one generation to the next.

Animals form attachments to objects (e.g. food, drink) that provide reinforcement (Tolman 1949; Skinner 1981; Garcia & Holder 1985), and positive experiences with foods encountered early in life can cause herbivores to prefer those foods later in life (reviewed by Provenza & Balph 1987, 1988). Herbivores can acquire attachments to different foods because behavioural (i.e. neurological, physiological and morphological) processes are not rigidly fixed genetically.

For instance, Distel and Provenza (1991) reared goats from 1 to 4 months of age with their mothers on blackbrush, a poorly digestible shrub high in condensed tannins and low in nitrogen (Provenza *et al.* 1983). These goats ingested 95% more blackbrush per unit body weight than goats naive to blackbrush immediately after the initial exposure of the former. Nine months later, after both groups of goats had foraged together on a grass-legume pasture, experienced goats still ingested 27% more blackbrush than inexperienced goats when *only* blackbrush was offered, and 30% more blackbrush when given a *choice* between alfalfa pellets and blackbrush. Physiological differences existed immediately following exposure. Goats reared on blackbrush excreted 63% more uronic acids per unit of body weight than inexperienced goats, perhaps because experienced goats were better able to detoxify depolymerization products from condensed tannins. There were also morphological differences immediately following exposure. The reticulo-rumen mass of goats reared on blackbrush was 30% greater than that of inexperienced goats. Thus, experience affected diet selection, and was related to neurological (preference), physiological (uronic acid), and morphological (rumen mass) changes.

The ability of herbivores to learn different foraging habits and skills (Flores *et al.* 1989a,b,c; Ortega-Reyes & Provenza 1993), means that many more foraging environments are suitable for survival than otherwise would be the case. This is consistent with observed variation in forage and habitat selection by different members of the same species. For example, sheep, cattle and bison (*Bison bison*), commonly considered grazers, often prefer and are productive on diets of shrubs (e.g. Zimmerman 1980; Provenza *et al.* 1983; Waggoner & Hinkes 1986). Likewise, goats and mule deer (*Odocoileus hemionus*), commonly considered browsers, often prefer and are productive on diets of grass (e.g. Willms &

McLean 1978; Austin & Urness 1983; Austin *et al.* 1983; Malechek & Provenza 1983; Urness *et al.* 1983).

When transgenerational links are broken, for example when wild and domesticated ruminants are introduced into new foraging environments, the animals are often less productive than conspecifics reared in that environment (reviewed by Griffith *et al.* 1989; Provenza & Balph 1990). Naive animals spend more time foraging but ingest less forage than experienced animals; they spend more time walking and walk for greater distances; they suffer more from predation and malnutrition, and they ingest more toxic plants (Provenza & Balph 1987, 1988; Provenza *et al.* 1992). In essence, the animals must learn about the new environment through trial and error, which is less efficient than learning from social models.

RETHINKING ADAPTATION

The foregoing empirical findings show that foraging involves post-ingestive and social feedback mechanisms. But empirical investigations have not usually been directed by a cogent behavioural–ecological theory. Ecological theories have been based primarily on functional models that relate the 'economic' properties of energy and nutrients to foraging. Foragers are supposedly pre-adapted to their environment and are thus unaffected by feedback mechanisms (but *see* Chapter 2). In what follows, we examine some of the reasons underlying the failure to reconcile empirical findings with theoretical treatments of foraging.

A history of models without histories

In the *Principia*, Newton demonstrated that immutable laws of motion governed all known physical objects. To calculate an object's trajectory, the history of the object was unimportant, and if physically unaltered by the experiment, the same object could be used to conduct the same experiment at any time without changing the outcome.

Biology, unlike physics, does not always mirror orderliness and predictability. Darwin and Wallace outlined dynamic rules that govern the existence of species. The past differed from the present, but how different and how far away from the present was it? Was adaptation so rapid that evolution could closely track the path of environmental fluctuations?

With the rediscovery of Mendel's work, it became clear to Darwinian biologists that genetic adaptations occur slowly. There was no Lamarckian demon, an organism that could quickly transmit environmentally induced modifications to its offspring. Thus, faith in the stability and generality of the present was maintained by the belief that evolution occurred gradually as a result of random variation and natural selection.

Neo-Darwinian biologists thus came to believe that adaptation through natural selection had driven organic design toward an equilibrium of niche-resident, co-existent, optimally adapted species (see critique by Gould & Lewontin 1979; cf. Mayr 1989). They believed that steady states, briefly interrupted by periods of adaptive transition, meant that equilibrium economic models which disregard historic continuity should adequately predict animal behaviour.

To seek an alternative to this paradigm is to assume that evolution has prepared animals to survive in a world that has always been in flux. Each individual, then, must adapt over a timescale much less than the generational time. But how? To explain, we consider the fundamentals of adaptation, first in the case of genetic evolution, and then as it pertains to learning.

Adaptation as a process

How can an organism adapt to an environment about which it lacks knowledge? Adaptation relies on three basic precepts. Firstly, it requires that organisms make errors (i.e. rules of variation, Staddon 1983, page 8). 'Error' making takes varied and complex forms, from genetic recombination and mutation to trial and error learning (e.g. sampling forages). Secondly, adaptation requires differential responses by organisms to a range of qualities (rules of selection, Staddon 1983). To these we add a third, which is the need for memory. Thus, if an organism at state A, with incomplete knowledge of present and future, is to transit to a qualitatively 'different' state B, it must:

1 Remain fairly stable in the present adaptive configuration, yet make enough 'errors' to explore alternatives without becoming extinct.

2 Retain true copies of 'errors' based on its response to the environment, thus entering a new transitional state from which it can again explore different alternatives.

3 Upon reaching state B, it must 'recognize' the adaptive value of B, and thus repeat step 1.

Allen (1990) discussed adaptation by natural selection as a process akin to 'hill climbing' (Fig. 5.1), where populations move toward elevated fitness once per generation. Differential response, or feedback, from the environment favours individuals of higher fitness, and thus the residual population appears to climb toward a 'peak of adaptation'. This process

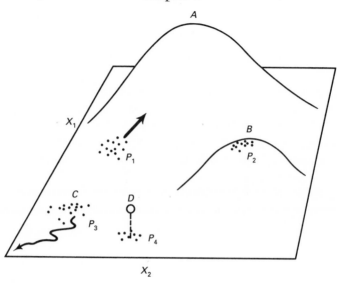

Fig. 5.1 Genetic evolution using the 'hill climbing' metaphor (adapted from Allen 1990). In this example, the process of adaptation is described by movement of the population through differential selection of deviates (by natural selection) on a 3-dimensional surface. The surface is composed (to keep the example within 3-dimensions) from only two physical characters (X_1, X_2) on horizontal axes, and a vertical axis (Y) representing fitness derived in the local environment. We have depicted four population behaviours. Two of these are *equilibrium behaviours* (associated with maxima) exemplified by hill-climbing dynamics: (i) population P_1 has entered the domain of attraction surrounding the global maximum (A) and is proceeding upwards, and (ii) population P_2 has moved to a position surrounding the equilibrium point, a local maximum (B). The two remaining behaviours are *non-equilibrium* in nature: (iii) the drift of population P_3, located on a flat region (C) and thus not under the influence of attractors (equilibria), and (iv) population P_4 has been driven to this spot by positive feedback created by sexual selection. In the latter case, sexual selection has imposed a source of fitness, shown as D, apart from that of the environmental fitness component associated with the combination of characters X_1 and X_2.

helps an organism find the tops of hills, and explore the 'topography' of a changing surface. However, there are hazards associated with limited knowledge of the terrain. Individuals may be located: (i) in a global maximum, (ii) in a local maximum, (iii) in a relatively flat region between global and local maxima, or (iv) in a positive feedback trap.

Locating the global maximum is analogous to climbing a mountain on a foggy day (Staddon 1983). The only feedback is the feeling of going upward, a cue that is lacking on a plateau, and thus may cause the climber to. wander in circles. On mountains with more than one peak, the climber will probably end up on a 'local maximum' rather than on the mountain top, the 'global maximum' (e.g. Andersen 1991).

Positive feedback traps are more complex. They represent a type of adaptive behaviour whose specific dynamics have been created as an artefact of the hill-climber's own adaptive mechanisms. In this case, the same mechanism that channels information from the local environment to the organism, and thus allows it to respond to environmental changes also generates its own 'environment of information'. This latter environment is distinct from the original object of adaptation, and may actually be maladaptive. For example, feedback traps in genetic variation include traits promoted by sexual selection, such as bright and unwieldy plumage or long courtship displays, that favour selection by prospective mates but increase the risk of predation.

The adaptationist paradigm (see critique by Gould & Lewontin 1979), which is the implicit basis for optimal foraging, assumes that animals are able to maintain themselves at or near the global maximum. As a result, models in optimal foraging theory (OFT) assume that innate laws of foraging lead to a diet that is optimal (maximize fitness) under environmental, physiological, and informational constraints (Schoener 1971; Janetos & Cole 1981).

Like an object in classical mechanics, OFT assumes that an individual's past is unimportant and ignores the steps to adaptation that include transition and memory. These assumptions permit researchers to apply analytical and predictive techniques from economics and engineering, such as classical optimization procedures (e.g. Charnov 1976) and optimization algorithms (e.g. Belovsky 1978; Mangel & Clark 1986).

A hierarchical model of adaptation

In contrast to the 'adaptationist' view, a hierarchical model (O'Neil *et al.* 1986) postulates: (i) genetic changes which operate in time units of 'generations', and (ii) learning within an individual's lifetime, which can transcend generations. Both genetic adaptation (evolution by generation) and learning (behavioural evolution by consequences) are products of natural selection. Both are based on selection rules that act on a range of variability. Evolution by generation results from selection involving genetic variability among individuals, while behavioural evolution within generations results from variation in the consequences of behavioural responses. Both cause change. For instance, evolution of preferences for forage items can occur through: (i) innate physiological variation among individuals in their responses to toxins (Keeler *et al.* 1988, 1992,

Provenza *et al.* 1992) accompanied by selection over many generations, presumably for genes controlling the synthesis of enzymes capable of detoxifying the compound (Provenza and Balph 1990); and (ii) behaviours acquired within individuals' lifetimes as a result of learning experiences early in life, and adoption of those behaviours by the individuals' offspring.

If we accept environmental change as an axiom, then Skinner (1981) identified the obvious importance of within-generational adaptation: 'Darwin and Spencer thought that selection would necessarily lead to perfection, but species, people, and cultures all perish when they cannot cope with rapid change.' Learning mechanisms that allow for rapid adaptation are consistent with natural selection. In variable environments (e.g. daily and weekly fluctuations in nutrient content and toxicity among and within plant species), flexible responses and rapid adaptation are primary correlates of survival; selection for finely tuned responses at an environmental steady state (e.g. 'optimal' diets) only proceeds secondarily.

Learning as adapting

Given the basic principles of adaptation (i.e. error-making, differential response and memory), the importance of learning by consequences becomes clear. Foragers must:
1 Recognize and ingest preferred food items while simultaneously sampling other items without endangering their future abilities to function and reproduce.
2 Alter their forage selection based on feedback.
3 Recognize a preferred state by remembering visual, olfactory, and gustatory cues, and thus repeat 1.

As with other hill-climbing processes, learning by consequence is highly adaptive, but not infallible (Skinner 1981; Staddon 1983). Each foraging experience has the potential to reshape diet selection, and as a result, foragers are prone to the pitfalls of perceived local maxima and positive feedback traps. By necessity, the process is highly sensitive to environmental perturbations. Thus, the diets of foragers may be considered transitional, even when preferences reflect the optimal diet. History, transition, and chance occurrence can create a variety of internal 'cultural states', as defined by their individual memory-resident assemblage of food preferences, even in organisms whose genetic backgrounds are similar.

TOWARDS A SIMULATION OF THE LEARNING
MODEL OF FORAGING

Theoretical ecologists have developed OFT models that assume steady-state behaviours satisfying explicit criteria (e.g. global fitness maxima and minima; Stephens & Krebs 1986), but these models do not address learning. Learning by consequences involves: (i) neurological, physiological and morphological mechanisms that allow adaptation and how they affect the ability to reach the global optimum, (ii) the history of the organism, and (iii) the consequences of temporal sequence, social and cultural variables that affect an individual's response to each food item. These feedback mechanisms facilitate the self-organization of diet selection (Fig. 5.2), and are the foundation of learning by consequences. Through learning by consequences, behaviour is shaped by the environment during the lifetime of an individual.

Objectives of a simulation

Our present work on a learning model of foraging (LMF) is an attempt to conceptualize more concisely the process of adaptation by consequences, and to illustrate the dynamics of that process (*see* Appendix for equations used in preliminary simulations). To be realistic, all of the major variables that influence dietary choice (the simulation output) must be included. Thus, for instance, the model should be able to simulate foragers that differ in past social experiences, age-related propensities for sampling new foods, and kind and amount of feedback received from ingesting different forages. Unless herbivores have similar foraging experiences, their use of forage items will probably differ.

Still in its early stages, LMF does not attempt to predict forage selection. Our approach is similar to theoretical explorations of classical OFT models (Emlen 1966; Charnov 1973; MacArthur & Pianka 1966; cf. Pyke *et al.* 1977), which differ from the explicit predictive goals of more

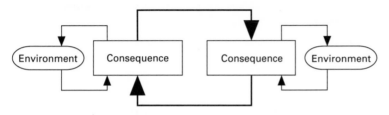

Fig. 5.2 Feedback loops that are active in adaptation by consequences.

recent formulations (Owen-Smith & Novellie 1982; Mangel & Clark 1988; Belovsky 1984). OFT models that attempt to predict diet selection assume that foraging is the result of evolutionary propensity to maximize the rate of assimilated energy, which can be expressed either as energy maximization or time minimization (Schoener 1971).

LMF is a hill-climbing model (hill climbing in reference to foraging choices has also been examined by Staddon 1980, 1983; Houston & McFarland 1980; and Sibly & McFarland 1976). Getting to the 'top of the hill' (i.e. maximizing energy intake) or climbing along the shortest path to a sufficient height (minimizing time spent foraging) are both possible outcomes, but LMF does not assume that foragers must find these routes. Ollason (1980) showed that a simple model of learning approximated predictions achieved by a model of optimal foraging in a patchy environment (Cowie 1977), based on Charnov's (1976) marginal value theorem.

The goal of LMF is to design a hill-climbing algorithm (i.e. an adaptive system), with limited information about the shape of the surface (hills and valleys), that describes the consequences of forage selection. Similar models have been used to explain foraging search patterns for ants (Goss *et al.* 1990; also discussed in Chapter 1), and the tendency to seek local optima and feedback traps in human economics (Allen & McGlade 1987a,b).

Hill climbing is based on past consequences of forage ingestion. Thus, the forager is at a greater disadvantage than the 'hill climber on a foggy day', because it cannot directly feel the slope of the hill as it climbs. Instead, it determines which direction to climb by memory of past experience (a relationship mediated by the senses of sight, smell and taste), and from additional feedback that occurs during feeding (similar to Ollason 1980). Thus, differences in past experiences are likely to cause genetically similar individuals to select different diets.

According to this model, herbivores may choose less nutritious diets when more nutritious choices are available (Fig. 5.3). They also may suffer from toxic overdoses when they have no prior experience with a toxic plant, and (or) if the lag time associated with positive post-ingestive consequences is shorter than the lag time associated with a toxic compound. Long time lags associated with some phytotoxins may result in toxicity, which is consistent with field and experimental observations (Provenza *et al.* 1992).

CONCLUSIONS

Foraging is a dynamic process adapted to environmental change. Economic models that assume forage selection resides at or near an optimum

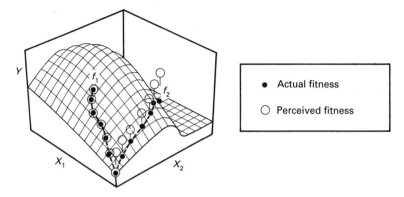

Fig. 5.3 A surface described by the fitness (Y) derived from ingesting two forages (X_1, X_2). In the learning model of foraging, choices are based on the forager's experiences, which differ among individuals with divergent foraging histories (f_1, f_2). Erroneous selections caused by long feedback time lags, or by conflicting feedbacks (e.g. rapid feedback from high energy content and long feedback lags from toxins in the same forage), may put foragers at risk of reducing their fitness (as in the final point of forager f_2).

do not consider the dynamic nature of this adaptive process and cannot explain why animals within a species express different dietary habits.

We believe that learning is vital to securing rapid adaptation. The learning model of foraging involves elements of history and change such as prior experiences, neuro- and morpho-physiological feedbacks and feedback lag times, social and cultural interactions, individual variability in responses to nutrients and toxins, forage sampling, and random events. Learning can activate a self-perpetuating series of consequences that are 'non-optimum', sensitive to perturbation, and difficult to predict.

Ecological theories have not incorporated learning as a within or between generational adaptive process, as is evident in their treatment of dynamic ecological processes as if they were events that reside at an optimum. In so doing, they ignore the basis for the very processes that make foraging dynamic. It is time for change: from describing events to understanding processes, and from predicting a unique endpoint to explaining means to disparate ends.

Explaining the dynamics of forage selection involves: (i) appreciation of the ecological implications of the uniqueness of individual animals, subgroups and cultures, each with its (their) own genetic and behavioural history, (ii) recognition that the consequences comprising each animal's foraging behaviours over time may not be stable, optimal, or predictable in the conventional sense, and (iii) study of foraging as a dynamic behavioural 'process' rather than a 'fixed' solution to a problem or as a static 'event'.

APPENDIX

The surface upon which learning occurs is defined by ingested quantities (X_1, X_2) of alternative forages and a potential response to their ingestion (Y, where $Y = f(X_1, X_2)$) that is ultimately a correlate of fitness. Thus, the three-dimensional surface corresponds to the future reproductive well-being of the animal due to digestible energy and nutrients, and accompanying non-nutritive plant compounds (e.g. toxins and indigestible fibre).

We assume that this surface is also correlated with neuronal impulses that signal positive and negative post-ingestive consequences, and that this feedback lags behind ingestion. Lag times may differ for each nutrient and chemical fraction. Lags may be correlated to rates of digestion, assimilation, and metabolism associated with each fraction, and may emanate from different innervated organs (rumen mucosa, liver, lower tract, Provenza *et al.* 1992). Simultaneously, nutrients and toxins are being cleared from the body by several processes (e.g. digestion, assimilation, detoxification and passage).

In the LMF simulation, the forager climbs to the highest fitness with limited *a priori* knowledge of the location of the 'peak of the hill'. This forager relies on all past experience (i.e. from *in utero* to present) and integrates feedbacks with these experiences.

A simulation model of this process employs a modification of the variable *relative attractivity* (A), a term originally defined in complex system decision models (Allen & McGlade 1987a,b), in which relative weights are assigned to each known alternative forage based on previous experiences of the herbivore. Consistent with the matching rule (Herrnstein 1970), the probability, P, of selecting a discrete choice i is equal to:

$$P_i = A_i / \Sigma_n A_n \qquad (5.1)$$

The attractivity of each alternative (A_0, the attractivity of *not* eating, is also an alternative) is reassessed at each discrete time step when there is one opportunity to ingest a 'bite' of forage. A_i is a function of two variables, utility U_i, and rationality, I,

$$A_i = e^{IU_i} \qquad (5.2)$$

In Equation 5.2, the variable I (where $0<I<1$) characterizes the tendency of a forager to deviate from 'rational' choices. When I approaches 0, attractivities of alternatives are similar, and the forager behaves as an 'adventurist', is less selective, and samples the unknown (Allen &

McGlade 1987a,b). At values of I near 1, differences among U_i create large differences between attractivities. In this case, the forager makes 'conservative' choices. In Allen and McGlade's (1987a,b) formulation of a human system, rationality is a constant. In our case, rationality values can be adjusted for food sampling and food neophobia, and there is evidence that I is a function of age (Provenza & Balph 1988; Mirza & Provenza 1992) and nutritional state (Provenza *et al.* 1992).

Let G be the forager's position on the fitness surface, Υ (i.e. the forager's perception of its condition). U_i is the effect on G which will be achieved by consuming another bite of alternative i (* denotes that the value of the variable relies on the condition that i is chosen):

$$U_i(t) = G^*(t+1) \tag{5.3}$$

The perceived position of the animal on the surface $\{X_1, X_2, G\}$ may or may not coincide with the actual surface $\{X_1, X_2, \Upsilon\}$, and misperceptions can affect forage selection. $G^\partial(t+1)$ is the average of weighted memories of feedbacks from having chosen i, δG^∂ (essentially, the partial derivative of G with respect to X_i, multiplied by the bite size of i),

$$G^*(t+1) = G(t) + \delta G^* \tag{5.4}$$

There are many possible algorithms for retaining past information (e.g. a running average). Simplistically, we weighted post-ingestive feedbacks, $\delta \Upsilon^\partial$ (discussed below) by grouping them in $\log_{10}$ memory cells. Each $\delta \Upsilon^\partial$ becomes a prior experience only after a lag time associated with that forage or forage fraction (e.g. energy, nutrient, toxin). Each memory cell, C_j, corresponds to the average of a range of prior experiences with choice i, $\delta \Upsilon^\partial_j$, beginning with the immediately preceding experience as $\delta \Upsilon^\partial_n$, and retracing to the first experience $\delta \Upsilon^\partial_1$. C_0 is the most recent feedback (i.e. the last $10^0 = 1$ experiences $\delta \Upsilon^\partial_n$); C_1 is the average of the last 10^1 experiences, not included in C_0 (i.e. $\delta \Upsilon^\partial_{n-1}$ to $\delta \Upsilon^\partial_{n-9}$), and likewise $C_2 = (\delta \Upsilon^\partial_{n-10} + \delta \Upsilon^\partial_{n-11} + \ldots + \delta \Upsilon^\partial_{n-99})/(10^2 - 10^1)$. These cells are averaged;

$$\delta G^\partial = (\Sigma_n C_n)/n \tag{5.5}$$

and thus recent experiences are more heavily weighted than past experiences. Post-ingestive feedbacks, $\delta \Upsilon^\partial_j$, received from the real surface $\{X_1, X_2, \Upsilon\}$ after a delay of time t_l, are augmented (Φ) by observational learning (e.g. a parental model).

$$\delta \Upsilon^*(t + t_l) = (\Upsilon(t) - \Upsilon(t-1))^\Phi = f(\delta X_i^*(t) - (\Sigma_n X_n e^{-r}_n)^\Phi) \tag{5.6}$$

At each time step, and for each forage choice, feedbacks are generated when choice X_i is added. Items in that pool are simultaneously depleted (i.e. assimilated, evacuated or detoxified) at the rate of r_i.

REFERENCES

Allen P.M. (1990) Why the future is not what it was: new models of evolution. *Futures* July/August.

Allen P.M. & McGlade J.M. (1987a) Evolutionary drive: The effect of microscopic diversity, error making and noise. *Foundations of Physics* **17**, 723–38.

Allen P.M. & McGlade J.M. (1987b) Modelling complex human systems: A fisheries example. *European J. Operational Res.* **30**, 147–67.

Andersen R. (1991) Habitat deterioration and the migratory behaviour of moose (*Alces alces* L.) in Norway. *J. Appl. Ecol.* **28**, 102–8.

Austin D.D. & Urness P.J. (1983) Overwinter forage selection by mule deer on seeded big sagebrush-grass range. *J. Wildl. Manage.* **47**, 1203–7.

Austin D.D., Urness P.J. & Fierro L.C. (1983) Spring livestock grazing affects crested wheatgrass regrowth and winter use by mule deer. *J. Range Manage.* **36**, 589–93.

Baker B.J., Booth D.A., Duggan J.P. & Gibson E.L. (1987) Protein appetite demonstrated: learned specificity of protein-cue preference to protein in adult rats. *Nutr. Res.* **7**, 481–7.

Baker B.J. & Booth D.A. (1989) Genuinely olfactory preferences conditioned by protein repletion. *Appetite* **13**, 223–7.

Bassette R., Fung D.Y.C. & Mantha V.R. (1986) Off-flavors in milk. *CRC Crit. Rev. Food Sci. Nutr.* **24**, 1–52.

Belovsky G.E. (1978) Diet optimization in a generalist herbivore: The moose. *Theor. Popul. Biol.* **14**, 105–34.

Belovsky G.E. (1984) Herbivore optimal foraging: A comparative test of three models. *Am. Nat.* **124**, 97–115.

Booth D.A. (1985) Food-conditioned eating preferences and aversions with introceptive elements: conditioned appetites and satieties. In *Experimental Assessments and Clinical Applications of Conditioned Food Aversions* (ed. by N.S. Braveman & P. Bronstein), pp. 22–41. New York Acad. Sci., New York.

Bond N. & DiGiusto E. (1975) Amount of solution drunk is a factor in the establishment of taste aversion. *Anim. Learn. Behav.* **3**, 81–4.

Bradley R.M. & Mistretta C.M. (1973) The gustatory sense in foetal sheep during the last third of gestation. *J Physiol.* **231**, 271–82.

Burritt E.A. & Provenza F.D. (1989a) Food aversion learning: ability of lambs to distinguish safe from harmful foods. *J. Anim. Sci.* **67**, 1732–9.

Burritt E.A. & Provenza F.D. (1989b) Food aversion learning: conditioning lambs to avoid a palatable shrub (*Cercocarpus montanus*). *J. Anim. Sci.* **67**, 650–3.

Burritt E.A. & Provenza F.D. (1990) Food aversion learning in sheep: persistence of conditioned taste aversions to palatable shrubs (*Cercocarpus montanus* and *Amelanchier alnifolia*). *J. Anim. Sci.* **68**, 1003–7.

Burritt E.A. & Provenza F.D. (1991) Ability of lambs to learn with a delay between food ingestion and consequences given meals containing novel and familiar foods. *Appl. Anim. Behav. Sci.* **32**, 179–89.

Burritt E.A. & Provenza F.D. (1992) Lambs form preferences for non-nutritive flavors paired with glucose. *J. Anim. Sci.* **70**, 1133–36.

Cannon D.S., Berman R.F., Baker, T.B. & Atkinson C.A. (1975) Effect of preconditioned stimulus experience on learned taste aversions. *J. Exp. Psychol.: Anim. Behav. Process.* **104**, 270–4.

Capretta P.J. & Rawls L.H. (1974) Establishment of a flavor preference in rats: importance of nursing and weaning experience. *J. Comp. Physiol. Psychol.* **88**, 670–3.

Cederlund G., Sandegren F. & Larsson K. (1987) Summer movement of female moose and dispersal of their offspring. *J Wildl. Manage.* **51**, 342–52.

Chapple R.S., Wodzicka-Tomaszewska M. & Lynch J.J. (1987a) The learning behavior of sheep when introduced to wheat. I. Wheat acceptance by sheep and the effect of trough familiarity. *Appl. Anim. Behav. Sci.* **18**, 157–62.

Chapple R.S., Wodzicka-Tomaszewska M. & Lynch J.J. (1987b) The learning behavior of sheep when introduced to wheat. II. Social transmission of wheat feeding and the role of the senses. *Appl. Anim. Behav. Sci.* **18**, 163–72.

Charnov E.L. (1973) Optimal foraging: Attack strategy of a mantid. *Am. Nat.* **110**, 141–51.

Charnov E.L. (1976) Optimal foraging: The marginal value theorem. *Theor. Popul. Biol.* **9**, 129–36.

Cowie R.J. (1977) Optimal foraging in great tits (*Parus major*). *Nature* **268**, 137–9.

Distel R.A. & Provenza F.D. (1991) Experience early in life affects voluntary intake of blackbrush by goats. *J. Chem. Ecol.* **17**, 431–50.

Domjan M. & Bowman T.G. (1974) Learned safety and the CS-US delay gradient in taste-aversion learning. *Learn. Motiv.* **5**, 409–23.

du Toit J.T., Provenza F.D. & Nastis A.S. (1991) Conditioned food aversion: how sick must a ruminant get before it detects toxicity in foods? *Appl. Anim. Behav. Sci.* **30**, 35–46.

Dragoin W.B. (1971) Conditioning and extinction of taste aversions with variations in intensity of the CS and the UCS in two strains of rats. *Psychon. Sci.* **22**, 303–5.

Egan A.R. & Rogers Q.R. (1978) Amino acid imbalance in ruminant lambs. *Aust. J. Agric. Res.* **29**, 1263–79.

Emlen J.M. (1966) The role of time and energy in food preference. *Am. Nat.* **100**, 611–17.

Festa-Bianchet M. (1986) Seasonal dispersion of overlapping mountain sheep ewe groups. *J. Wildl. Manage.* **50**, 325–30.

Flores E.R., Provenza F.D. & Balph D.F. (1989a) Role of experience in the development of foraging skills of lambs browsing the shrub serviceberry. *Appl. Anim. Behav. Sci.* **23**, 271–8.

Flores E.R., Provenza F.D. & Balph D.F. (1989b) The effect of experience on the foraging skills of lambs: Importance of plant form. *Appl. Anim. Behav. Sci.* **23**, 285–91.

Flores E.R., Provenza F.D. & Balph D.F. (1989c) Relationship between plant maturity and foraging experience of lambs grazing hycrest crested wheatgrass. *Appl. Anim. Behav. Sci.* **23**, 279–84.

Galef B.G. Jr. (1985a) Direct and indirect behavioral pathways to the social transmission of food avoidance. In *Experimental Assessments and Clinical Applications of Conditioned Food Aversions* (ed. by N.S. Braveman & P. Bronstein), pp. 203–15. New York Acad. Sci., New York.

Galef B.G. Jr. (1985b) Socially induced diet preference can partially reverse a LiCl-induced diet aversion. *Anim. Learn. Behav.* **13**, 415–18.

Galef B.G. Jr. (1986) Social interaction modifies learned aversions, sodium appetite, and both palatability and handling-time induced dietary preference in rats (*Rattus norvegicus*). *J. Comp. Psychol.* **4**, 432–9.

Galef B.G. Jr. & Beck M. (1990) Diet selection and poison avoidance by mammals individually and in social groups. In *Handbook of Behavioral Neurobiology, Vol. 10: Neurobiology of Food and Fluid Intake* (ed. by E.M. Stricker) pp. 329–49. Plenum Press, New York.

Galef B.G. Jr. & Sherry D.F. (1973) Mother's milk: a medium for transmission of cues reflecting the flavor of mother's diet. *J. Comp. Psychol.* **83**, 374–8.

Garcia J. (1989) Food for Tolman: Cognition and cathexis in concert. In *Aversion, Avoidance and Anxiety* (ed. by T. Archer & L. Nilsson) pp. 45–85. Hillsdale, N.J.

Garcia J. & Holder M.D. (1985) Time, space and value. *Human Neurobiol.* **4**, 81–9.

Garcia J., Ervin F.R., Yorke C.H. & Koelling R.A. (1967) Conditioning with delayed vitamin injections. *Science* **155**, 716–18.

Gasaway W.C., DuBois S.D., Boertje R.D., Reed D.J. & Simpson D.T. (1989) Response of radio-collared moose to a large burn in central Alaska. *Can. J. Zool.* **67**, 325–9.

Gibson E.L. & Booth D.A. (1989) Dependence of carbohydrate-conditioned flavor preference on internal state in rats. *Learn. Motiv.* **20**, 36–47.

Goss S., Beckers R., Deneubourg J.L., Aron S. & Pasteels J.M. (1990) How trail laying and trail following can solve foraging problems for ant colonies. In *Behavioral Mechanisms of Food Selection*, (ed. by R.N. Hughes) *NATO ASI Series, vol. G 20* pp. 661–78. Springer Verlag, Berlin.

Gould S.J. & Lewontin R.C. (1979) The spandrels of San Marco and the Panglossian paradigm: A critique of the adaptationist programme. *Proc. Roy. Soc. London B* **205**, 581–98.

Green G.C., Elwin R.L., Mottershead B.E. & Lynch J.J. (1984) Long-term effects of early experience to supplementary feeding in sheep. *Proc. Aust. Soc. Anim. Prod.* **15**, 373–5.

Green K.F. & Garcia J. (1971) Recuperation from illness: flavor enhancement in rats. *Science* **173**, 749–51.

Griffith B., Scott J.M., Carpenter J.W. & Reed C. (1989) Translocation as a species conservation tool: Status and strategy. *Science* **245**, 477–80.

Grovum W.L. (1988) Appetite, palatability and control of food intake. In *The Ruminant Animal: Digestive Physiology and Nutrition* (ed. by D.C. Church), pp. 202–6. Prentice Hall, Englewood Cliffs.

Hepper P.G. (1988) Adaptive fetal learning: prenatal exposure to garlic affects postnatal preferences. *Anim. Behav.* **36**, 935–6.

Herrnstein R.J. (1970) On the law of effect. *J. Expt. Analy. Behav.* **13**, 243–66.

Hill D.L. & Przekop P.R. Jr. (1988) Influences of dietary sodium on functional taste receptor development: a sensitive period. *Science* **241**, 1826–8.

Hill D.L. & Mistretta C.M. (1990) Developmental neurobiology of salt taste sensation. *TINS* **13**, 188–95.

Hinch G.N., Lecrivain E., Lynch J.J. & Elwin R.L. (1987) Changes in maternal–young associations with increasing age of lambs. *Appl. Anim. Behav. Sci.* **17**, 305–18.

Houston A.I. & McFarland D.J. (1980) Behavioral resilience and its relation to demand functions. In *Limits to Action: The Allocation of Individual Behavior* (ed. by J.E.R. Staddon), pp. 177–205. Academic Press, New York.

Hunter R.F. & Milner C. (1963) The behavior of individual, related and groups of south country Cheviot hill sheep. *Anim. Behav.* **11**, 507–13.

Janetos A.C. & Cole B.J. (1981) Imperfectly optimal animals. *Behav. Ecol. Sociobiol.* **9**, 203–10.

Kalat J.W. & Rozin P. (1973) 'Learned-Safety' as a mechanism in long-delay taste-aversion learning in rats. *J. Comp. Physiol.* **83**, 198–207.

Keeler R.F., Baker D.C. & Evans J.O. (1988) Individual animal susceptibility and its relationship to induced adaptation or tolerance in sheep to *Galea officinalis*. L. *Vet. Hum. Toxicol.* **30**, 420–3.

Keeler R.F., Baker D.C. & Panter K.E. (1992) Concentration of galegine in *Verbesina enceliodes* and *Galega officinalis* and the toxic and pathologic effects induced by the plants. *J. Environ. Path. Toxic. Oncol.* **11**, 11–17.

Key C. & MacIver R.M. (1980) The effects of maternal influences on sheep: breed differences in grazing, resting and courtship behavior. *Appl. Anim. Ethol.* **6**, 33–48.

Lane M.A., Ralphs M.A., Olsen J.D., Provenza F.D. & Pfister J.A. (1990) Conditioned taste aversion: potential for reducing cattle loss to larkspur. *J. Range Manage.* **43**, 127–31.

Launchbaugh K.L. & Provenza F.D. (1993) Can plants practice mimicry to avoid grazing by mammalian herbivores? *Oikos* (in press).

Launchbaugh K.L., Provenza F.D. & Burritt E.A. (1993) How herbivores track variable environments: response to variability of phytotoxins. *J. Chem. Ecol.* (in press).

Lobato J.F.P., Pearce G.R. & Beilharz R.G. (1980) Effect of early familiarization with dietary supplements on the subsequent ingestion of molasses-urea blocks by sheep. *Appl. Anim. Ethol.* **6**, 149–61.

Lynch J.J. (1987) The transmission from generation to generation in sheep of the learned behaviour for eating grain supplements. *Aust. Vet. J.* **64**, 291–2.

Lynch J.J., Keogh R.G., Elwin R.L., Green G.C. & Mottershead B.E. (1983) Effects of early experience on the post-weaning acceptance of whole grain wheat by fine-wool Merino lambs. *Anim. Prod.* **36**, 175–83.

MacArthur R.H. & Pianka E. (1966) On the optimal use of a patchy environment. *Am. Nat.* **100**, 603–9.

Malechek J.C. & Provenza F.D. (1983) Feeding behavior and nutrition of goats on rangelands. *World Anim. Rev.* **47**, 38–48.

Mangel J. & Clark C.W. (1986) Towards a unified foraging theory. *Ecology.* **67**, 1127–38.

Mangel M. & Clark C.W. (1988) *Dynamic Modeling in Behavioral Ecology.* Princeton University Press, Princeton, N.J.

Mayr E. (1989) *Toward a New Philosophy of Biology: Observations of an Evolutionist.* Harvard University Press, Cambridge.

Mehiel R. & Bolles R.C. (1984) Learned flavor preferences based on caloric outcome. *Anim. Learn. Behav.* **12**, 421–7.

Mehiel R. & Bolles R.C. (1988) Learned flavor preferences based on calories are independent of initial hedonic value. *Anim. Learn. Behav.* **16**, 383–7.

Mirza S.N. & Provenza F.D. (1990) Preference of the mother affects selection and avoidance of foods by lambs differing in age. *Appl. Anim. Behav. Sci.* **28**, 255–63.

Mirza S.N. & Provenza F.D. (1992) Effects of age and conditions of exposure on maternally mediated food selection in lambs. *Appl. Anim. Behav. Sci.* **33**, 35–42.

Morrill J.L. & Dayton A.D. (1978) Effect of feed flavor in milk and calf starter on feed consumption and growth. *J. Dairy Sci.* **61**, 229–32.

Nolte D.L. & Provenza F.D. (1991) Food preferences in lambs after exposure to flavors in milk. *Appl. Anim. Behav. Sci.* **32**, 381–389.

Nolte D.L., Provenza F.D. & Balph D.F. (1990) The establishment and persistence of food preferences in lambs exposed to selected foods. *J. Anim. Sci.* **68**, 998–1002.

Ollason J.G. (1980) Learning to forage – optimally? *Theo. Pop. Biol.* **18**, 44–56.

O'Neil R.V., DeAngelis D.L., Waide J.B. & Allen T.F.H. (1986) *A Hierarchical Concept of Ecosystems. Monographs in Population-Biology, 23.* Princeton University Press, Princeton, N.J.

Ortega-Reyes L. & Provenza F.D. (1993) Amount of experience and age affect the development of foraging skills of goats browsing blackbrush (*Caleogyne romosissima*). *Appl. Anim. Behav. Sci.* (in press).

Owen-Smith N. & Novellie P. (1982) What should a clever ungulate eat? *Am. Nat.* **119**, 151–78.

Panter K.E. & James L.F. (1990) Natural plant toxicants in milk: a review. *J. Anim. Sci.* **68**, 892–904.

Pfister J.A., Provenza F.D. & Manners G.D. (1990) Ingestion of tall larkspur by cattle: separating the effects of flavor from post-ingestive consequences. *J. Chem. Ecol.* **16**, 1697–1705.

Provenza F.D. (1977) *Biological manipulation of blackbrush* (Coleogyne ramosissima *Torr.*) *by browsing with goats*. Thesis, Utah State Univ., Logan.

Provenza F.D. & Balph D.F. (1987) Diet learning by domestic ruminants: theory, evidence and practical implications. *Appl. Anim. Behav. Sci.* **18**, 211–32.

Provenza F.D. & Balph D.F. (1988) Development of dietary choice in livestock on rangelands and its implications for management. *J. Anim. Sci.* **66**, 2356–68.

Provenza F.D. & Balph D.F. (1990) Applicability of five diet-selection models to various foraging challenges ruminants encounter. In *Behavioral Mechanisms of Food Selection*, (ed. by R.N. Hughes) *NATO ASI Series, vol. G 20* pp. 423–59. Springer Verlag, Berlin.

Provenza F.D. & Burritt E.A. (1991) Socially induced diet preference ameliorates conditioned food aversion in lambs. *Appl. Anim. Behav. Sci.* **31**, 229–36.

Provenza F.D., Pfister J.A. & Cheney C.D. (1992) Mechanisms of learning in diet selection with reference to phytotoxicosis in herbivores. *J. Range Manage.* **45**, 36–45.

Provenza F.D., Lynch J.V., Burritt E.A. & Scott C.B. (1993). How goats learn to distinguish between novel foods that differ in postingestive consequences. Oikos (submitted).

Provenza F.D., Bowns J.E., Urness P.J., Malechek J.C. & Butcher J.E. (1983) Biological manipulation of blackbrush by goat browsing. *J. Range Manage.* **36**, 513–18.

Provenza F.D., Burritt E.A., Clausen T.P., Bryant J.P., Reichardt P.B. & Distel R.A. (1990) Conditioned flavor aversion: a mechanism for goats to avoid condensed tannins in blackbrush. *Am. Nat.* **136**, 810–28.

Pyke G.H., Pulliam H.R. & Charnov E.L. (1977) Optimal foraging: A selective review of theory and tests. *Quart. Rev. Biol.* **52**, 137–54.

Ralphs M.H. & Olsen J.D. (1990) Adverse influence of social facilitation and learning context in training cattle to avoid eating larkspur. *J. Anim. Sci.* **68**, 1944–52.

Roath L.R. & Krueger W.C. (1982) Cattle grazing and behavior on forested range. *J. Range Manage.* **35**, 332–8.

Rogers Q.R. & Egan A.R. (1975) Amino acid imbalance in the liquid-fed lamb. *Aust. J. Biol. Sci.* **28**, 169–81.

Rozin P. & Kalat J.W. (1971) Specific hungers and poison avoidance as adaptive specializations of learning. *Psychol. Rev.* **78**, 459–86.

Rozin P. (1976) The selection of foods by rats, humans and other animals. In *Advances in the Study of Behavior* (ed. by J.S. Rosenblatt, R.A. Hinde, E. Shaw & C. Beer), pp. 21–76. Academic Press, New York.

Schoener T.W. (1971) Theory of feeding strategies. *Annu. Rev. Ecol. Syst.* **2**, 369–403.

Sibly R.M. & McFarland D.J. (1976) On the fitness of behavior sequences. *Am. Nat.* **110**, 601–17.

Skinner B.F. (1981) Selection by consequences. *Science* **213**, 501–4.

Smotherman W.P. (1982) Odor aversion learning by the rat fetus. *Physiol. Behav.* **29**, 769–71.

Squibb R.C., Provenza F.D. & Balph D.F. (1990) Effect of age of exposure on consumption of a shrub by sheep. *J. Anim. Sci.* **68**, 987–97.

Staddon J.E.R. (1980) Optimality analyses of operant behavior and their relation to optimal foraging. In *Limits to Action: The Allocation of Individual Behavior* (ed. by J.E.R. Staddon), pp. 101–42. Academic Press, New York.

Staddon J.E.R. (1983) *Adaptive Behavior and Learning.* Cambridge University Press, New York.

Stephens D.W. & Krebs J.R. (1986) *Foraging Theory.* Princeton University Press, Princeton, N.J.

Stickrod G., Kimble D.P. & Smotherman W.P. (1982) In utero taste/odor aversion conditioning in the rat. *Physiol. Behav.* **28**, 5–7.

Thorhallsdottir A.G., Provenza F.D. & Balph D.F. (1987) Food aversion learning in lambs with or without a mother: discrimination, novelty and persistence. *Appl. Anim. Behav. Sci.* **18**, 327–40.

Thorhallsdottir A.G., Provenza F.D. & Balph D.F. (1990a) Ability of lambs to learn about novel foods while observing or participating with social models. *Appl. Anim. Behav. Sci.* **25**, 25–33.

Thorhallsdottir A.G., Provenza F.D. & Balph D.F. (1990b) The role of the mother in the intake of harmful foods by lambs. *Appl. Anim. Behav. Sci.* **25**, 35–44.

Thorhallsdottir A.G., Provenza F.D. & Balph D.F. (1990c) Social influences on conditioned food aversions in sheep. *Appl. Anim. Behav. Sci.* **25**, 45–50.

Tolman E.C. (1949) There is more than one kind of learning. *Psychol. Rev.* **56**, 144–55.

Urness P.J., Austin D.D. & Fierro L.C. (1983) Nutritional value of crested wheatgrass for wintering mule deer. *J. Range Manage.* **36**, 225–6.

Waggoner V. & Hinkes M. (1986) Summer and fall browse utilization by an Alaskan bison herd. *J. Wildl. Manage.* **50**, 322–4.

Williams R.J. (1978) You are extraordinary. In *The Art of Living* (ed. by R.R. Leichtmann & C. Japibase), pp. 121–3. Berkeley Books, New York.

Willms W. & McLean A. (1978) Spring forage selected by tame mule deer on big sagebrush range, British Columbia. *J. Range Manage.* **31**, 192–9.

Zahorik D.M. & Houpt K.A. (1991) Species differences in feeding strategies, food hazards, and the ability to learn food aversions. In *Foraging Behavior* (ed. by A.C. Kamil & T.D. Sargent), pp. 289–310. Garland, New York.

Zahorik D.M., Maier S.F. & Pies R.W. (1974) Preferences for tastes paired with recovery from thiamine deficiency in rats: appetitive conditioning or learned safety? *J. Comp. Physiol. Psychol.* **87**, 1083–91.

Zimmerman E.A. (1980) Desert ranching in central Nevada. *Rangelands* **2**, 184–6.

6: Hunger-Dependent Diet Selection in Suspension-Feeding Zooplankton

WILLIAM R. DEMOTT

When considering the potential for behavioural flexibility in diet selection, it is useful to classify consumers into two groups depending on whether food items are captured and processed individually or in bulk (Hughes 1980). Microphages collect and process numerous small particles in bulk and thus have little opportunity for actively selecting between individual food items, whereas macrophages capture food items individually and therefore have much greater opportunity for decision-based food selection. Zooplankton are often described as 'filter feeders', a term which suggests a passive, microphagous mode of feeding. For a filter feeder, food selection is largely a function of the range of particle sizes which can be retained by the feeding apparatus and successfully ingested (Boyd 1976). Within the past 15 years, however, numerous studies have shown that many taxa of suspension-feeding zooplankton are not filter-feeding microphages but are able to use complex behaviours to select between individual particles which differ in size or nutritional value (reviewed by Price 1988; DeMott 1990; Vanderploeg 1990). Even among primarily microphagous taxa, complex mechanisms, including behavioural flexibility, may influence particle selection near the lower and upper size limits imposed by the feeding apparatus.

New findings on the food-selection capabilities of zooplankton naturally raise questions about how these capabilities are utilized in nature. This chapter is based on the premise that the most effective approach to studying feeding behaviour combines an understanding of physiology and functional morphology with theoretical models based on fitness (*see also* Chapters 3 and 8). Optimal diet models predict how food selection can maximize nutritional gains in homogeneous patches. Most tests of optimal diet models have used carnivores or granivores, organisms whose prey are high in nutritional quality (reviewed by Stephens Krebs 1986). For these organisms, diet selection usually is a compromise between small prey that can be rapidly captured and consumed and large prey that require longer handling times but provide more energy per item (e.g. Werner & Hall 1974). Suspension-feeding zooplankton, however, live in environments in which nutritious algae are mixed with particles of low food value, such as silt, detritus, and digestion-resistant or toxic algae (Porter 1977). In this situation an ability to select nutritious particles should be advantageous, provided that the costs of handling and rejecting individual particles are low.

Unlike conventional optimal diet models, theoretical models for suspension-feeding zooplankton have assumed that pre-ingestion handling time is negligible and that gut processing places upper limits on ingestion rates (Lehman 1976; Taghon 1981). It follows from these assumptions that nutritional benefit can be maximized by food selection based on particle quality rather than particle size. Lehman's (1976) optimal diet model assumes that the 'filter feeder' first captures particles and then rejects or ingests individual items after assessing their nutritional value. If rejection costs are not significant, acceptance of low-quality particles depends solely on the abundance of high-quality particles. The proportion of low-quality particles in diet should decline to zero when the abundance of better particles is sufficient to saturate gut processing. Thus, in common with conventional optimal diet models, Lehman's (1976) model predicts that discrimination against low-ranking foods should be strong when preferred foods are abundant and weak when preferred foods are scarce. These two classes of models can be distinguished by subjecting animals to rapid changes in food concentration which disrupt the normal equilibrium between hunger and food availability.

This chapter examines diet selection in two of the best-studied zooplankton taxa: calanoid copepods, which are primarily macrophages, and cladocerans of the genus *Daphnia*, which are primarily microphages. Feeding mechanisms and possible modes of selectivity are reviewed, and then behavioural responses to variation in food concentration and hunger. Experimental results are used to examine the importance of hunger in modulating diet choice in these suspension-feeders.

Mechanisms of particle capture and selectivity in calanoid copepods

The development of high-speed filming techniques (200–500 frames/ sec) in the late 1970s revealed that the term 'filter feeder' grossly underrates the sensory and particle-handling capabilities of calanoid copepods (reviewed by Koehl 1984; Price 1988). Two modes of particle capture have been observed: raptorial capture of large particles and passive capture of small particles. Raptorial feeding requires first the detection of the particle in the feeding current and then an active response. High-speed films have revealed that calanoid copepods can detect individual particles at a distance and may orient their body position or alter feeding currents to increase the probability of capture (e.g. Strickler 1982). Larger algae are detected at greater distances and are more likely to elicit precapture responses. Thus, preferences for larger algae may simply reflect an increased probability of detection, a perceptual bias, rather than a decision-based process (Price & Paffenhöfer 1985).

Decision-based selection ('active choice') appears to be important in selection based on nutritional quality, rather than particle size. High-speed movies show that captured particles are manipulated and often appear to be tasted at the mouth before being ingested or rejected. Low-quality particles, including fecal pellets, dead algae, toxic algae and plastic beads are more likely to be rejected after capture than are nutritious, living algae (e.g. Paffenhöfer & Van Sant 1985; Vanderploeg *et al.* 1990).

Complementing the filming studies, laboratory experiments, mostly conducted with pairs of particles, have shown that calanoid copepods have impressive abilities to discriminate between particles which differ in nutritional value. Food-selection experiments have shown that calanoids can discriminate between algal-flavoured and unflavoured artificial particles (Poulet & Marsot 1978; DeMott 1986), between algae and microspheres of similar size (reviewed by DeMott 1988a), between live and dead algae of the same species (Paffenhöfer & Van Sant 1985; DeMott 1988b), between digestible and digestion-resistant algae (DeMott 1989), between toxic and non-toxic algae (Huntley *et al.* 1986; DeMott & Moxter 1991) and even between nitrogen-sufficient and nitrogen-limited algae of the same species (Cowles *et al.* 1988; Butler *et al.* 1989).

Despite the food selection capabilities of calanoids, analyses of the gut contents and fecal pellets from copepods feeding in natural waters provide evidence of rather broad diets, often including substantial amounts of detritus, silt, blue-green algae and other low-quality particles (Turner 1984; Hart 1987). Some experiments in which copepods fed in natural seston provide evidence of weak or non-existent selection between algal species (Huntley 1981; Turner & Tester 1989) and relatively weak discrimination between algae and artificial particles (Gliwicz 1977; Bern 1990a). These studies raise the possibility that copepods feeding in natural seston are less selective than animals feeding on only pairs of particles. Perhaps the wide variety of phytoplankton and detritus food in natural waters can overload the sensory or particle-handling capabilities of copepods (cf. Milinski 1990).

One alternative explanation for the apparent conflict between field and laboratory results is that a copepod's feeding selectivity depends on the animal's level of hunger. Laboratory experiments and, especially, cinema-graphic studies, typically use relatively high food concentrations, whereas zooplankton are often food-limited in nature. In agreement with optimal diet theory (Lehman 1976), food concentration-variable selectivity has recently been documented in laboratory experiments pairing algae and algal-flavoured microspheres (DeMott 1988a), live and dead algae (DeMott 1988b, 1989), readily-digested and digestion-resistant algae (DeMott 1989), and green algae and non-toxic cyanobacteria (DeMott & Moxter 1991). Thus, attempts to characterize copepod feeding selectivity need to consider the effects of food availability and hunger on performance.

Contrary to Lehman's (1976) diet model, concentration-dependent selectivity has also been documented for diaptomid copepods in experiments in which readily-ingested green flagellates with large diatoms (Vanderploeg *et al.* 1988) were paired with a large, spined green alga (DeMott 1990). These results suggest that handling costs or handling times could be significant for large, difficult-to-handle algae. Handling times as short as 0.06 sec have been reported for copepods feeding on readily-ingested algae (Gill & Poulet 1988). Plastic spheres, which were usually rejected, required handling times of 0.3 sec for smaller spheres (10–25 μm) and 'several seconds' for larger spheres. Thus, 'recognition time' (Hughes 1979) could be important for low-quality particles. In contrast to the extremely rapid ingestion of small algae, Vanderploeg *et al.* (1988) reported that *Diaptomus sicilis* required 2.2 sec to ingest a 120-μm-long segment of the filamentous diatom *Melosira islandica*. Filming studies have also shown that several particles can be handled simultaneously and that additional particles can be actively captured and ingested while others continue to be manipulated (Strickler 1984; Van-derploeg & Paffenhöfer 1985). Thus, quantifying handling times may have limited utility for evaluating diet models for copepods.

Mechanisms of particle capture and feeding selectivity in *Daphnia*

Daphnids possess large combs on the third and fourth limbs which are usually described as 'filter combs'. Although some workers have argued that the 'filter combs' act as solid paddles (Gerritsen *et al.* 1988), detailed studies of morphology and function indicate that water is sucked through the filter combs as the space between adjacent limbs widens (Fryer 1991). The intersetular spacing ('mesh-size') of the filter combs varies between species, with body size, and, to a lesser extent, along the filters of individuals (reviewed by Lampert 1987). Measurements of mesh-size in *Daphnia* range from 0.24–1.8 µm with most values <1 µm. The 'sieving hypothesis' of food collection gains further support from experiments showing a good correlation between filter mesh-size and selectivity for very small particles (e.g. DeMott 1985; Brendelberger 1991). Particles collected on the filter limbs are brushed dorsally into the food groove and then pushed anteriorly to the mouth.

As expected for a microphage, daphnids exhibit relatively non-selective feeding over a broad size range of particles with reduced 'retention efficiency' as the lower and upper size limits are approached (reviewed by Lampert 1987). Unlike copepods, daphnids do not discriminate between algae and inert particles (DeMott 1986) or between toxic and non-toxic algae (e.g. Lampert 1981) within this middle size range. There is, however, some evidence for behavioural flexibility near the upper size limits. For example, Gliwicz and Siedlar (1980) found that *Daphnia* can respond to the presence of filamentous blue-green algae by narrowing the gap between the valves of the carapace. Although this behaviour reduces feeding rates for small particles it is thought to be more effective in preventing the entry of large filaments that interfere with feeding and may be toxic. In *in situ* feeding experiments with plastic beads, Gliwicz (1977) documented a seasonal decline in the size range of beads ingested, which was correlated with the abundance of filamentous algae. He did not, however, test for changes in size selectivity in controlled experiments.

When daphnids fed in experimental mixtures of lake water and plastic beads, Gliwicz (1980) and Bern (1990b) found that larger beads were common in the food groove but less common in the gut. This finding implies that larger beads were captured but were rejected at the mouth. Bern (1990b) also found that large spherical algae were 'preferred' over beads of the same size. Although these preferences may indicate selection based on food quality, the algae, unlike plastic beads, could have been deformed or crushed during ingestion.

Hartmann and Kunkel (1991) allowed *Daphnia pulicaria* to feed on mixtures of *Oscillatoria* filaments and small algae, rapidly anaesthetized the animals, and then observed the positions of collected particles using scanning electron microscopy. They observed substantial separation of filaments and small cells within the food groove. This result suggests that the post-capture rejection of masses of particles by *Daphnia* can be more selective than previously thought.

One might predict that hungry daphnids, like hungry copepods, would be less likely to reject filaments and other particles near the upper size limit for ingestion. Despite the popularity of *Daphnia* for feeding studies (reviewed by Lampert 1987), the possibility of hunger-dependent food selection on large particles has not been critically tested. In one possible example, Meise *et al.* (1985) observed variable 'preferences' for small *Chlamydomonas minutissima* (3–8 μm) over large *Chlamydomonas capensis* (4–40 μm).

TERMS AND METHODS

Zooplankton feeding rates are usually expressed as clearance rates and ingestion rates. The term 'clearance rate' which is synonymous with 'filtering rate' in much of the literature, is a measure of the volume of water from which particles have been removed per unit time (e.g. ml ind^{-1} h^{-1}). Ingestion rates are measures of the biomass or volume of food eaten and can be calculated by multiplying the clearance rate by the food concentration. To facilitate comparisons, all food concentrations are reported here in volumetric units, where 1 ppm (part per million) = 10^6 μm^3 ml^{-1}.

Equivalent clearance rates on different classes of food particles imply non-selective feeding and comparisons of clearance rates provide a direct measure of the magnitude of feeding selectivity. For this reason, most of the feeding-rate data presented in this chapter are expressed as clearance rates. Food selection is quantified by the selectivity coefficient a (Chesson 1983). All experiments involved pairs of particles and particle depletion was negligible, due to short duration. In this simple case, a is the ratio of the clearance rate on one food to the sum of the clearance rates on both food types. The index ranges from 0 to 1.0 with a value of 0.50 indicating non-selective feeding. One-way ANOVA was used to determine whether selectivity varied in response to food concentration or hunger.

The food-selection experiments presented here were conducted at two laboratories using zooplankton collected in nearby lakes: the Max Planck Institute of Limnology, Plön, Germany; and the Crooked Lake Biological Station, near Columbia City, Indiana. Experiments run in Germany used mixtures of two very similar species of copepod, *Eudiaptomus gracilis* and *Eudiaptomus graciloides* (see DeMott 1989).

All experiments used the radiotracer method for quantifying feeding rates. In this method, animals are first acclimated with unlabelled food particles and then allowed to feed in a suspension containing radioactively labelled food particles for a period shorter than the gut-passage time (5–15 min). Clearance rates are estimated by comparing the radioactivity of the animals (dpm/ind) with the radioactivity of the feeding suspension (dpm/ml). Selection between pairs of food types is quantified using experiments in which one food type is labelled with [14C]bicarbonate and the other with [32P]orthophosphate (see DeMott 1988a, 1989 for further details, including methods for labelling algae with isotopes and using liquid scintillation counting to assess radioactivity).

Three variants of the radiotracer method employed differing acclimation conditions. Typically, the unlabelled suspensions used for acclimating animals and the labelled suspensions used for quantifying feeding rates contain the same combinations and concentrations of food particles. However, high sensitivity and short feeding times make the radiotracer technique highly suitable for testing the effects of hunger on feeding rates and diet selection behaviour. In this second variant, animals are acclimated to high concentrations of unlabelled particles and feeding behaviour is tested after various periods without food; or animals are starved and then feeding behaviour is tested after various periods with food.

The third variant of the radiotracer technique involves acclimating animals to particular food conditions and then adding low concentrations of labelled particles directly to the feeding suspension. This variant provides a means for investigating the feeding behaviour of animals acclimated to the complex mixtures of algae, detritus, and inert particles found in the seston of natural waters. Effects of hunger and food concentration can be examined by comparisons with animals acclimated with filtered water (starved), natural seston diluted with filtered water, or natural seston enriched with algae from laboratory cultures.

HUNGER-DEPENDENT FEEDING BEHAVIOUR

Effects of food concentration and hunger on ingestion rates

As is typical of most predators, the ingestion rate of suspension-feeding zooplankton increases with increasing food concentration until reaching a plateau at high concentrations. The beginning of the plateau is often termed the 'incipient limiting concentration' (ILC; McMahon & Rigler 1965). Typically, the clearance rate is maximal and constant below the ILC and the ingestion rate increases more or less linearly with increasing food concentration (see review in Lampert 1987). Some marine copepods, however, decrease their clearance rates at very low food concentrations (Frost 1975). This 'feeding threshold' has been interpreted as a reduced feeding effort when the potential energy gain is less than the energy expended for feeding.

Above the ILC the ingestion rate is constant and maximal and the clearance rate declines in proportion to increases in food concentration. An early study with *Daphnia* using foods ranging from bacteria to protozoa showed that the ILC was determined by food volume and not particle number (McMahon & Rigler 1965). In experiments with *Daphnia* feeding on a wide variety of algae, Geller (1975) found that the particle volume, taking into account gut packing, was a better predictor of the ILC than was carbon content. On the other hand, Libourel Houde and Roman (1987) found that the marine copepod *Acartia* compensated for food with low protein content by ingesting greater volumes. Thus, it seems likely that ILC is determined by both physical gut packing and biochemical correlates of hunger and satiation.

As mentioned earlier, foraging models for suspension-feeding zooplankton are based on the assumption that gut processing and hunger, rather than pre-ingestion handling, place limits on the maximal ingestion rate. This assumption receives strong support from experiments which test the effects of rapid changes in food concentration on the functional response. A recent study by Lampert *et al.* (1988) confirms earlier observations (e.g. McMahon & Rigler 1965; Geller 1975) and provides details on time scales. Figure 6.1 shows the functional response curves for *Daphnia* which were acclimated to experimental concentrations of a green alga for 24 h (solid circles) or starved for 24 h (open circles) and then allowed to feed for 5 to 10 min on radioactively labelled cells. Animals which had been acclimated to experimental food concentrations exhibited a typical functional response with a plateau (ILC) beginning at about 2 ppm. In contrast, starved animals exhibited no evidence of a plateau, even at order-of-magnitude higher food concentrations.

Although the experiment presented in Fig. 6.1 used an acclimation of 24 h, time course experiments show that zooplankton respond much more quickly to marked changes in food concentration. For example, when previously starved *Daphnia* were transferred to a high concentration of the green alga *Scenedesmus* (13 ppm), the starvation effect was

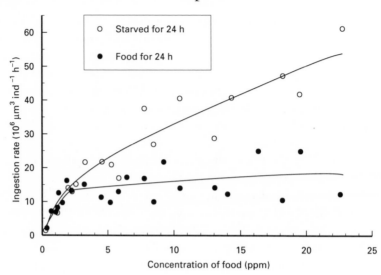

Fig. 6.1 Effects of starvation on the functional response of *Daphnia pulicaria* (2.8 mm length) feeding on *Scenedesmus acutus* (after Lampert *et al.* 1988). Daphnids were acclimated to experimental concentrations of food for 24 h (solid circles) or starved for 24 h (open circles). Each data point represents the mean of 7–9 individuals.

reduced within 4 min and the equilibrium ingestion rate was approached within 30 min (Lampert *et al.* 1988). When animals were transferred from the high food concentration to filtered lake water, increased feeding rates were observed after a few minutes of starvation and the full response was reached after about 3 h. A second series of experiments tested the responses of *Daphnia longispina* and a copepod, *Eudiaptomus gracilis*, to less extreme changes in food concentration (alternations between 0.20 and 2.0 ppm). Both species exhibited very similar responses and, as might be expected, the reduced range of food concentration resulted in a slower approach to equilibrium following transfer to the high food level (about 2 h) and a somewhat shorter time to the full starvation response following transfer to the low food concentration (about 1.5 h).

How can these time-course experiments be interpreted? The initial responses to marked changes in food concentration are on the same time scale as gut filling (5–10 min at high food concentrations). The longer term responses (0.5–3 h) are probably associated with biochemical correlates of satiation and hunger. A further response, on a time scale of days, has been documented in marine copepods (e.g. Landry & Hassett 1985; Hassett & Landry 1990) but has not been investigated in freshwater zooplankton. Increases in food concentration and changes in food quality can lead to the induction of digestive enzymes which influence assimilation efficiency and maximal ingestion rates. Zooplankton also exhibit reduced clearance rates following prolonged (1–4 days) starvation (Muck & Lampert 1980; Hassett & Landry 1990), presumably due to weakened condition and reduced digestive capabilities.

Effects of food concentration and hunger on food selection by copepods

According to the optimal diet model for suspension-feeding zooplankton (Lehman 1976), behaviourally flexible taxa should exhibit hunger-dependent food selection between high- and low-quality food particles. In a recent study, the freshwater calanoid copepod *Eudiaptomus* spp. exhibited shifts in selectivity that agree with the general, qualitative predictions of the optimal diet model (DeMott 1989). Low quality particles, including digestion-resistant algae and dead algae, were weakly discriminated against at low food concentrations and strongly discriminated against at high food concentrations. The relation between the functional response and selectivity was examined in experiments pairing a high-quality green flagellate, *Chlamydomonas reinhardi* (6 μm diameter) with equivalent concentrations (by volume) of a slightly larger (10 μm), green alga, *Crucigenia tetrapedia*. *Crucigenia* is covered by a gelatinous sheath that reduces its digestibility and its nutritional value per volume (Porter 1976). Copepods fed non-selectively then offered both algae together at low concentration, but became selective as the maximal ingestion rate was approached and exhibited strong (4:1) selection against *Crucigenia* at the two highest concentrations (Fig. 6.2; ANOVA, shift in selectivity, $p < 10^{-6}$). *Daphnia pulicaria* feeding within the same beakers showed evidence of saturated feeding, yet fed non-selectively at all food concentrations (DeMott 1989).

A second, previously unpublished experiment, differentiates the effects of hunger and food concentration in these shifts in selectivity. When starved for 16 h and then fed high-concentration mixtures (2.0 ppm each) of *Chlamydomonas* and *Crucigenia*, *Eudiaptomus* spp. initially fed non-selectively but showed increasing selection against *Crucigenia* during the first hour of acclimation to the high density of food (Fig 6.3, *see* page 113; ANOVA, change in selectivity, $p < 0.001$). There was also a small difference in selectivity between high- and low-concentration mixtures offered without pre-feeding (time = 0 treatments). A similar, rapid shift in selectivity was observed when the copepod *Diaptomus birgei* was

Chapter 6

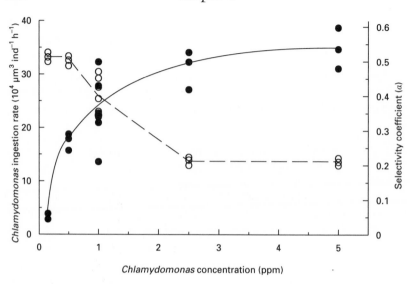

Fig. 6.2 Relation between food concentration, ingestion rate, and selectivity for *Eudiaptomus* spp. feeding on equal volume mixtures of *Chlamydomonas reinhardi* and *Crucigenia tetrapedia*. Open circles are selectivity coefficients; solid circles are ingestion rates on *Chlamydomonas*. Each circle is the mean value for an experimental beaker (after DeMott 1989).

starved for 24 h and then allowed to feed on high-concentration mixtures of *Chlamydomonas* and a larger, dead flagellate, *Carteria* sp. (DeMott 1990). Experiments have also been conducted in which *Diaptomus birgei* was first acclimated to high- concentration mixtures of *Chlamydomonas* and low-quality foods and then allowed to feed on low concentrations of labelled particles after periods of starvation (0–24 h). These experiments with dead *Carteria* (DeMott 1990) and a non-toxic blue-green alga (DeMott & Moxter 1991) show a gradual relaxation in selection against low-quality particles over the first 6 h of starvation.

Experiments by the author on the effects of hunger on copepod feeding selectivity complement the work discussed above concerning the effects of hunger on the functional responses of *Daphnia* and *Eudiaptomus* (Lampert *et al.* 1988). In both series of experiments, starved animals responded very quickly to sudden increases in food whereas well-fed animals responded more gradually to transfer to a food-free environment. *Daphnia*'s more rapid response to starvation in comparison to calanoid copepods (3 h vs. 6 h for full response) may reflect *Daphnia*'s higher metabolic rate and greater sensitivity (Muck & Lampert 1984).

Natural seston typically contains dozens of species of algae and numerous non-living particles, which differ widely in size, shape and

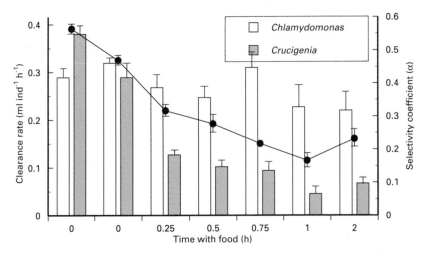

Fig. 6.3 Time course for the starvation effect on the selective feeding of *Eudiaptomus* spp. on mixtures of *Chlamydomonas* and *Crucigenia*. The first and second (time = 0) treatments represent trials with low and high concentrations, respectively, of labelled algae (see text for further explanation). Data are means ± 1 SE for two replicated beakers per treatment.

nutritional value. Can laboratory experiments with pairs of food particles predict copepod feeding selectivity in complex mixtures of natural foods? This question is addressed in the following preliminary experiment which involved adding low concentrations of dual-labelled *Chlamydomonas* and *Crucigenia* to copepods feeding in lake water. A range of food concentrations was created by using water from two lakes in the vicinity of the Max Planck Institute of Limnology and by including treatments with filtered lake water (starved) and lake water enriched with 2.0 ppm *Chlamydomonas*. Phytoplankton concentrations in lake water were estimated by counting and measuring with an inverted microscope. One lake, Schöhsee, had a low concentration of phytoplankton dominated by small flagellates whereas the other lake, Kellersee, had a very high abundance of phytoplankton dominated by the large flagellate *Cryptomonas ovata*. *Eudiaptomus* spp. collected from Schöhsee were acclimated to experimental conditions for 3–4 h before the addition of dual-labelled *Chlamydomonas* and *Crucigenia*.

Copepods acclimated to natural lake seston discriminated more strongly against digestion-resistant *Crucigenia* at higher food concentrations (Fig. 6.4). The range of selectivity values and the effects of concentration are virtually identical to results from simple laboratory mixtures (compare Figs. 6.2 and 6.4). Thus, when the effects of food concentration and hunger are quantified, experiments with pairs of particles can

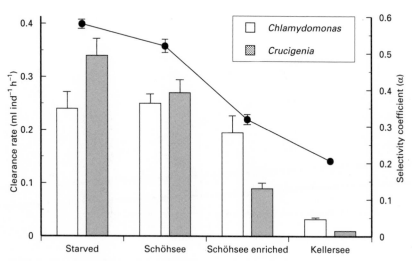

Fig 6.4 Selective feeding by *Eudiaptomus* spp. on mixtures of *Chlamydomonas* and *Crucigenia*. The copepods fed in beakers of filtered lake water ('starved'), seston from a lake with low phytoplankton abundance ('Schöhsee'; algal volume, 0.6 ppm), seston enriched with 2.0 ppm of *Chlamydomonas* ('Schöhsee enriched') and seston from a lake with a high phytoplankton abundance ('Kellersee'; algal volume, 5.1 ppm). Data are means ± 1 SE for three replicate beakers per treatment.

accurately predict copepod feeding selectivity in complex mixtures of natural seston. Copepods feeding in Kellersee water had full, dark guts indicative of intensive feeding on *Cryptomonas*. The very low clearance rates in Kellersee water probably reflects a food concentration far above the incipient limiting concentration. Gelatinous algae, including *Crucigenia*, were not detected in the natural seston from either lake. Since the animals were exposed to low concentrations of labelled algae for only 10 min, any learned avoidance of *Crucigenia* must occur very quickly.

Assuming that hunger underlies shifts in selectivity, one can predict that factors which inhibit feeding would lead to reduced discrimination against low-quality particles. This prediction is supported by an experiment which examined the effect of a high abundance of a predatory *Cyclops* on the feeding rates and selectivity of *Eudiaptomus* spp. In the absence of the predator, *Eudiaptomus* spp. exhibited shifts in selectivity between *Chlamydomonas* and *Crucigenia* (Table 6.1), similar to results described above. In trials run a few days later under the same food conditions, the presence of predators strongly inhibited feeding on *Chlamydomonas* and the presumably hungry copepods did not discriminate against *Crucigenia*. Although the density of predatory *Cyclops* was

Table 6.1 Effects of food concentration and the presence of predatory *Cyclops vicinus* (about 50 l⁻¹) on the clearance rates and selectivity of *Eudiaptomus* spp. for *Chlamydomonas reinhardi* and *Crucigenia*. Copepods were acclimated to filtered water ('starved'), seston with a relatively low food concentration ('Schöhsee'; algal volume, 0.6 ppm), and seston enriched with 2.0 ppm of *C. reinhardi*. Data are means ± 1 SE for 3 replicate beakers per treatment.

Acclimation treatment	Clearance rate (ml ind⁻¹ h⁻¹)		Selectivity coefficient (a)
	Chlamydomonas	*Crucigenia*	
Without *Cyclops*			
Starved	0.24 ± 0.029	0.340 ± 0.026	0.58 ± 0.017
Schöhsee	0.25 ± 0.019	0.150 ± 0.018	0.37 ± 0.023
Enriched	0.20 ± 0.031	0.093 ± 0.013	0.32 ± 0.021
With *Cyclops* (50 l⁻¹)			
Starved	0.035 ± 0.010	0.064 ± 0.008	0.66 ± 0.036
Schöhsee	0.062 ± 0.003	0.110 ± 0.011	0.64 ± 0.013
Enriched	0.060 ± 0.004	0.120 ± 0.007	0.68 ± 0.012

exceptionally high, other studies have shown that realistic densities of predatory copepods influence swimming behaviour (Ramcharan & Sprules 1991) and inhibit feeding (Folt & Goldman 1981) in suspension-feeding diaptomids (*see* Chapter 9 for a general discussion of predation risk).

Contrary to Lehman's (1976) optimal diet model, diaptomid copepods show concentration-dependent discrimination against presumably high-quality but difficult-to-ingest algae (Vanderploeg *et al.* 1988; DeMott 1990). Vanderploeg *et al.* (1988) hypothesized that a high abundance of difficult-to-handle particles exhausted the copepods' motivation or ability to carry out the complex manoeuvres required for successful ingestion. To help separate the effects of hunger from handling costs, adult female *Diaptomus birgei* were acclimated to a wide range of concentrations of *Chlamydomonas* and then low-concentration mixtures (0.25 ppm each) of dual-labelled *Chlamydomonas* and *Fragilaria crotenensis* were added for 10-min feeding trials. The large, ribbon-like diatom *Fragilaria crotenensis* was isolated from Crooked Lake, where it is the most common diatom during summer stratification. Analysis of the gut contents of *D. birgei* from the lake indicates that colonies of *Fragilaria* (mean 50 μm × 300 μm) are dismembered into small pieces (on the order of 30 μm × 6 μm) as they are ingested. *Fragilaria* was cleared at a very high rate when the abundance of a readily ingested alga (*Chlamy-*

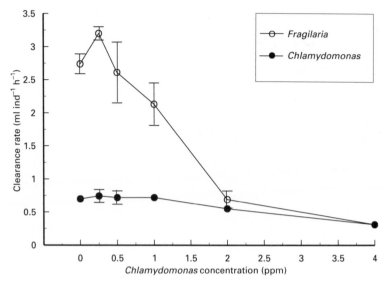

Fig. 6.5 Effects of *Chlamydomonas* concentration on selection by adult female *Diaptomus birgei* between *Chlamydomonas* and *Fragilaria crotenensis*. Copepods were acclimated for 3–4 h on unlabelled *Chlamydomonas* before the addition of a low-concentration mixture (0.5 ppm total) of dual labelled cells. Data are means ± 1 SE for three replicate beakers per treatment.

domonas) was low, but was largely rejected when *Chlamydomonas* was abundant (Fig. 6.5). Thus, food concentration and, presumably, hunger, can strongly influence selection for difficult-to-handle particles, even when these particles are present in low abundance.

Recent filming studies have shown that large, poor quality particles, including toxic filamentous cyanobacteria and nylon rods, are captured and handled before being rejected (Vanderploeg *et al.* 1990). The effects of handling difficulties might, therefore, be separated from satiation effects by testing whether a high abundance of large, consistently rejected particles influences selectivity for large particles of good nutritional quality. Two series of experiments were run in Crooked Lake during late summer, when high quality algae were scarce but *Aphanizomenon*, a filamentous cyanobacterium which is strongly rejected by copepods (DeMott & Moxter 1991) was abundant. The presumably hungry copepods feeding in unenriched lake seston exhibited selection against *Fragilaria* but not against the gelatinous green alga, *Pandorina* (Fig. 6.6). These results provide preliminary support for a 'handling costs' hypothesis, such as the one proposed by Vanderploeg *et al.* (1990). Although more carefully controlled experiments are needed, the results shown in

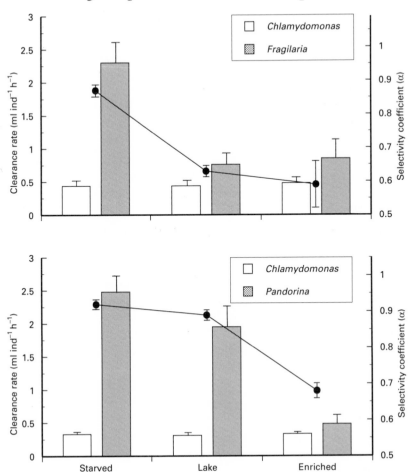

Fig. 6.6 Effects of pre-feeding conditions on selection between *Chlamydomonas* and *Fragilaria* or *Pandorina* by adult female *Diaptomus birgei*. Copepods were acclimated for 3–4 h in filtered lake water ('starved'), natural seston ('lake'), or natural seston enriched with 2.0 ppm of *Chlamydomonas* ('enriched') before the addition of low-concentration mixtures (0.5 ppm total) of dual labelled cells. Data are means ± 1 SE for three replicate beakers per treatment.

Figs. 6.5 and 6.6 suggest that both hunger and handling costs, separately or together, can influence selection for large, difficult-to-handle algae.

Effects of food concentration and hunger on selective feeding by *Daphnia*

Although daphnids have limited ability to handle individual food items they appear to have some ability to regulate the ingestion of large particles, including filaments. The following preliminary experiment was designed to test for behavioural flexibility in feeding on mixtures of *Chlamydomonas* and a filamentous cyanobacterium, *Oscillatoria agardhii* (mean filament length 800 μm). This strain from Lake Washington is toxic to *Daphnia* (Infante & Abella 1985). *Daphnia pulicaria* from Crooked Lake were cultured in the laboratory on a diet of *Chlamydomonas*. Daphnids were then placed in filtered water ('starved') or acclimated with moderately high concentrations (2.0 ppm) of *Chlamydomonas* or *Oscillatoria* alone. After 3–4 h of acclimation, a low-concentration mixture of dual-labelled *Chlamydomonas* and *Oscillatoria* (0.25 ppm each) was added and animals were allowed to feed for 10 min.

Pre-feeding conditions strongly influenced both selectivity and clearance rates (Fig. 6.7; ANOVA, selectivity and clearance rates, both $p < 0.001$). Starved *Daphnia* fed non-selectively on the two food types whereas *Daphnia* acclimated to *Chlamydomonas* or *Oscillatoria* exhibited selection against *Oscillatoria*. These results support Gliwicz's (1977) hypothesis that exposure to filaments leads to both selection against filaments and a general inhibition of feeding in *Daphnia*. Acclimation to high levels of high-quality foods, however, also favoured

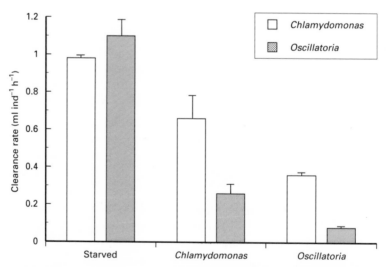

Fig. 6.7 Effects of pre-feeding conditions on selection between *Chlamydomonas* and *Oscillatoria agardhii* by juvenile *Daphnia pulicaria* (1.2 mm length). Animals were starved or acclimated with moderately high (2.0 ppm) concentrations of *Chlamydomonas* or *Oscillatoria* alone before the addition of low-concentration mixtures of dual-labelled cells. Data are means ± 1 SE for three replicate beakers per treatment.

selection against *Oscillatoria*. Microscopic observations revealed that *Daphnia* acclimated to *Chlamydomonas* had full guts whereas *Daphnia* acclimated in *Oscillatoria* had relatively empty guts. Thus, differing levels of hunger produced relatively similar results. Since other studies have shown non-selective feeding on filaments, even at relatively high concentrations (e.g. Lampert 1981; Fulton 1988), further experiments are needed to verify these results and to examine the possible effects of filament toxicity.

CONCLUSIONS

Data presented and reviewed here illustrate a substantial degree of behavioural flexibility in diet selection by suspension-feeding zooplankton. When combined with a knowledge of proximate feeding mechanisms, an optimal diet model provided a guide to the design of the experiments described and a useful basis for generalizing about diet selection in suspension-feeding zooplankton. In general agreement with the predictions of the optimal diet model (Lehman 1976), low-quality particles are selected against more strongly when high-quality particles are abundant. In agreement with the assumptions of the model, hunger, rather than food concentration *per se*, controls food-selection behaviour. However, the abundance of difficult-to-handle particles also seems to influence selection, even for hungry animals.

Excellent agreement was found between selectivity experiments with copepods acclimated to pairs of cultured algae and experiments with the same taxa of copepods and labelled algae, but with acclimation to complex mixtures of natural seston. Thus, when the effects of food concentration and hunger are quantified, experiments with pairs of particles can accurately predict diet selection behaviour in complex mixtures of natural seston. Rejection of low-quality particles by copepods appears to be highly stereotyped.

As is often the case in dealing with the complexities of nature, the concept of 'macrophage versus microphage' appears to describe the opposite ends of a continuum, rather than a distinct dichotomy. Even primarily microphagous taxa, such as *Daphnia*, can exhibit behavioural flexibility in diet selection. The now discredited view of copepods as 'filter feeders' was based, in part, on an underestimation of sensory capabilities, particle handling skills, and potential behavioural complexity in 'lower' animals. Algae and other food particles were considered too small and too abundant to be handled individually. The error in this reasoning has been re-exposed by recent studies showing that suspension-feeding protozoa can discriminate between live and dead bacteria (Landry *et al.* 1991) and can exhibit concentration-dependent feeding on bacteria and bacteria-sized microspheres (Nygaard *et al.* 1988). Thus, even the simplest consumers can exhibit relatively complex behaviours and even the smallest prey have predators which can capture and ingest them individually.

Strong links between gut fulness, hunger and diet-selection behaviour are certainly not limited to suspension-feeding zooplankton. Studies with carnivores as well as many terrestrial herbivores (*see* reviews in Hughes (1990) and also Chapters 3 and 8) indicate that gut-processing constraints are often important when food resources are abundant and differ in digestibility or nutritional adequacy. The behaviour of small fish feeding on suspension-feeding zooplankton offers a number of parallels with zooplankton feeding behaviour. Hungry fish prey size-selectively on large *Daphnia*, in a manner that maximizes energy gained per unit handling time (Werner & Hall 1974). After an extended feeding period, however, the not-so-hungry fish feed selectively on smaller, more difficult-to-catch copepods which are easier to digest and contain more lipids (Confer & O'Bryan 1989). Much can be learned about diet selection behaviour by manipulating hunger.

REFERENCES

Bern L. (1990a) Size-related discrimination of nutritive and inert particles by freshwater zooplankton. *J. Plankton Res.* **12**, 1059–67.

Bern L. (1990b) Post-capture particle size selection by *Daphnia cucullata* (Cladocera). *Limnol. Oceanogr.* **35**, 923–6.

Boyd C.M. (1976) Selection of particle sizes by filter-feeding copepods: A plea for reason. *Limnol. Oceanogr.* **21**, 175–80.

Brendelberger H. (1991) The filter-mesh size of cladocerans predicts retention efficiency for bacteria. *Limnol. Oceanogr.* **36**, 884–94.

Butler N.M., Suttle C.A. & Neill W.E. (1989) Discrimination by freshwater zooplankton between single algal cells differing in nutritional status. *Oecologia* **78**, 368–72.

Chesson J. (1983) The estimation and analysis of preference and its relationship to foraging models. *Ecology* **65**, 1297–304.

Confer J. & O'Bryan L.M. (1989) Changes in prey rank and preference by young planktivores for short-term and long-term ingestion periods. *Can. J. Fish. Aquatic Sci.* **46**, 1026–32.

Cowles T.J., Olson R.J. & Chisholm S.W. (1988) Food selection by copepods: discrimination on the basis of food quality. *Mar. Biol.* **100**, 41–9.

DeMott W.R. (1985) Relations between filter mesh-size, feeding mode, and capture efficiency for cladocerans feeding on ultrafine particles. *Arch. Hydrobiol. Beih. Ergeb. Limnol.* **21**, 125–34.

DeMott W.R. (1986) The role of taste in food selection by freshwater zooplankton. *Oecologia* **69**, 334–40.

DeMott W.R. (1988a) Discrimination between algae and artificial particles by freshwater and marine copepods. *Limnol. Oceanogr.* **33**, 397–408.

DeMott W.R. (1988b) Discrimination between algae and detritus by freshwater and marine zooplankton. *Bull. Mar. Sci.* **43**, 486–99.

DeMott W.R. (1989) Optimal foraging theory as a predictor of chemically mediated food selection by suspension feeding copepods. *Limnol. Oceanogr.* **34**, 140–54.

DeMott W.R. (1990) Retention efficiency, perceptual bias, and active choice as mechanisms of food selection by suspension-feeding zooplankton. In *Behavioural Mechanisms of Food Selection*, (ed. by R.N. Hughes), *NATO ASI series, vol. G 20*, pp. 569–94. Springer Verlag, Berlin.

DeMott W.R. & Moxter F. (1991) Foraging on cyanobacteria by copepods: responses to chemical defenses and resource abundance. *Ecology* **72**, 1820–34.

Folt C.L. & Goldman C.R. (1981) Allelopathy between zooplankton: a mechanism for interference competition. *Science* **213**, 1133–5.

Frost B.W. (1975) A threshold feeding behavior in *Calanus pacificus*. *Limnol. Oceanogr.* **20**, 259–62.

Fryer G. (1991) Functional morphology and the adaptive radiation of the Daphniidae (Brachiopoda: Anomopoda). *Phil. Trans. R. Soc. London B* **331**, 1–99.

Fulton R.S. (1988) Grazing on filamentous algae by herbivorous zooplankton. *Freshwater Biol.* **20**, 263–72.

Geller W. (1975) Die Nahrungsaufnahme von *Daphnia pulex* in Abhängigkeit von der Futterkonzentration, die Temperatur, der Körpergrösse und dem Hungerzustand der Teire. *Arch. Hydrobiol. Suppl.* **48**, 47–107.

Gerritsen J., Porter K.G. & Strickler J.R. (1988) Not by sieving alone: observations of suspension feeding in *Daphnia*. *Bull. Mar. Sci.* **43**, 377–94.

Gill C.W. & Poulet S.A. (1988) Impedance traces of copepod appendage movements illustrating sensory feeding behaviour. *Hydrobiologia* **167/168**, 303–10.

Gliwicz Z.M. (1977) Food size selection and seasonal succession of filter feeding zooplankton in an eutrophic lake. *Ekol. Pol. A* **17**, 663–708.

Gliwicz Z.M. (1980) Filtering rates, food size selection, and feeding rates in cladocerans—another aspect of interspecific competition in filter-feeding zooplankton. In *Evolution and Ecology of Zooplankton Communities* (ed. by W.C. Kerfoot), pp. 282–91. University Press of New England, Hanover, N.H.

Gliwicz Z.M. & Siedlar E. (1980) Food size limitation and algae interfering with food collection in *Daphnia*. *Arch. Hydrobiol.* **88**, 155–77.

Hart R. (1987) Observations on calanoid diet, seston, phytoplankton—zooplankton relationships, and inferences on calanoid food limitation in a silt-laden reservoir. *Arch. Hydrobiol.* **111**, 67–82.

Hartmann H.J. & Kunkel D.D. (1991) Mechanisms of food selection in *Daphnia*. *Hydrobiologia* **225**, 129–54.

Hassett R.P. & Landry M.R. (1990) Effects of diet and starvation on digestive enzyme activity and feeding behaviour of the marine copepod *Calanus pacificus*. *J. Plankton Res.* **12**, 991–1010.

Hughes R.N. (1979) Optimal diets under the energy maximization premise: the effects of recognition time and learning. *Am. Nat.* **113**, 209–21.

Hughes R.N. (1980) Optimal foraging in the marine context. *Oceanogr. Mar. Biol. Ann. Rev.* **18**, 428–49.

Hughes R.N. (1990) *Behavioural Mechanisms of Food Selection,* (ed. by R.N. Hughes) *NATO ASI series, vol. G 20.* Springer Verlag, Berlin.

Huntley M. (1981) Nonselective, nonsaturated feeding by three calanoid copepod species in the Labrador Sea. *Limnol. Oceanogr.* **26**, 831–42.

Huntley M., Sykes P., Rohan S. & Marin V. (1986) Chemically-mediated rejection of dinoflagellate prey by the copepods *Calanus pacificus* and *Paracalanus parvus:* mechanism, occurrence, and significance. *Mar. Ecol. Prog. Ser.* **28**, 105–20.

Infante A. & Abella S.E.B. (1985) Inhibition of *Daphnia* by *Oscillatoria* in Lake Washington. *Limnol. Oceanogr.* **30**, 1046–52.

Koehl M.A.R. (1984) Mechanisms of particle capture by copepods at low Reynolds numbers: possible modes of selective feeding. In *Trophic Interactions within Aquatic Ecosystems AAAS Select. Symp. Ser. 85* (ed. by D.G. Meyers & J.R. Strickler), pp. 135–66. Westview, Boulder, CO.

Lampert W. (1981) Inhibitory and toxic effects of blue-green algae on *Daphnia. Int. Rev. ges. Hydrobiol.* **66**, 285–98.

Lampert W. (1987) Feeding and nutrition in *Daphnia.* In *Daphnia* (ed. by R.H. Peters & R. de Bernardi). *Mem. 1st. Ital. Idrobiol.* **45**, 143–92.

Lampert W., Schmitt R.-D. & Muck P. (1988) Vertical migration of freshwater zooplankton: test of some hypotheses predicting a metabolic advantage. *Bull. Mar. Sci.* **43**, 620–40.

Landry M.R. & Hassett R.P. (1985) Time scales in behavioral, biochemical, and energetic adaptations to food-limiting conditions by a marine copepod. *Arch. Hydrobiol. Veih. Ergeb. Limnol.* **21**, 209–22.

Landry M.R., Lehnerfournier J.M., Sundstrom J.A., Fagernes V.L. & Selph K.E. (1991) Discrimination between living and heat-killed prey by a marine zooflagellate *Paraphysomona vestita. J. Exp. Mar. Biol. Ecol.* **146**, 139–51.

Lehman J.T. (1976) The filter feeder as an optimal forager, and the predicted shapes of feeding curves. *Limnol. Oceanogr.* **21**, 501–16.

Libourel Houde S.E. & Roman M.R. (1987) Effects of food quality on the functional ingestion response of the copepod *Acartia tonsa. Mar. Ecol. Progr. Ser.* **40**, 69–77.

McMahon J.W. & Rigler R.H. (1965) Feeding rate of *Daphnia magna* Strauss in different foods labeled with radioactive phosphorus. *Limnol. Oceanogr.* **10**, 105–13.

Meise C.J., Munns Jr. W.R. & Hairston Jr. N.G. (1985) An analysis of the feeding behavior of *Daphnia pulex. Limnol. Oceanogr.* **30**, 862–70.

Milinski M. (1990) Information overload and food selection. In *Behavioural Mechanisms of Food Selection,* (ed. by R.N. Hughes) *NATO ASI series, vol. G 20,* pp. 721–36. Springer Verlag, Berlin.

Muck P. & Lampert W. (1980) Feeding of freshwater filter-feeders at very low food concentrations: Poor evidence for 'threshold feeding' and 'optimal foraging' in *Daphnia longispina* and *Eudiaptomus gracilis. J. Plankton Res.* **2**, 367–79.

Muck P. & Lampert W. (1984) An experimental study on the importance of food conditions for the relative abundance of calanoid copepods and cladocerans. *Arch. Hydrobiol./Suppl.* **66**, 157–79.

Nygaard K., Borsheim K.Y. & Thingstad T.F. (1988) Grazing rates on bacteria by marine heterotrophic microflagellates compared to uptake of bacteria-sized monodisperse fluorescent latex beads. *Mar. Ecol. Prog. Ser.* **44**, 159–65.

Paffenhöfer G.-A. & Van Sant K.B. (1985) The feeding response of a marine planktonic copepod to quantity and quality of particles. *Mar. Ecol. Prog. Ser.* **27**, 55–65.

Porter K.G. (1976) Enhancement of algal growth and productivity by grazing zooplankton. *Science* **192**, 1332–4.

Porter K.G. (1977) The plant–animal interface in freshwater ecosystems. *Am. Sci.* **65**, 159–70.

Poulet S.A. & Marsot P. (1978) Chemosensory grazing by marine calanoid copepods (Arthropoda: Crustacea). *Science* **200**, 1403–5.

Price H.J. (1988) Feeding mechanisms in marine and freshwater zooplankton. *Bull. Mar. Sci.* **43**, 327–43.

Price H.J. & Paffenhöfer G.-A. (1985) Perception of food availability by calanoid copepods. *Arch. Hydrobiol. Beih. Ergeb. Limnol.* **21**, 115–24.

Ramcharan C.W. & Sprules W.G. (1991) Predator-induced behavioral defense and its consequences for two calanoid copepods. *Oecologia* **86**, 276–86.

Stephens D.W. & Krebs J.R. (1986) *Foraging Theory*. Princeton, N.J. University Press, Princeton, New Jersey.

Strickler J.R. (1982) Calanoid copepods, feeding currents, and the role of gravity. *Science* **218**, 158–60.

Strickler J.R. (1984) Sticky water: a selective force in copepod evolution. In *Trophic Interactions within Aquatic Ecosystems AAAS Select. Symp. Ser. 85* (ed. by D.G. Meyers & J.R. Strickler), pp. 135–66. Westview, Boulder, CO.

Taghon G.L. (1981) Beyond selection: optimal ingestion rate as a function of food value. *Am. Nat.* **118**, 202–14.

Turner J.T. (1984) Zooplankton feeding ecology: content of fecal pellets of the copepods *Eucalanus pileatus* and *Paracalanus quasimodo* from continental shelf water of the Gulf of Mexico. *Mar. Ecol. Prog. Ser.* **15**, 27–46.

Turner J.T. & Tester P.A. (1989) Zooplankton feeding ecology: nonselective feeding by copepods *Acartia tonsa* Dana, *Centropages velificatus* De Oiveira, and *Eucalanus pileatus* Giesbrecht in the plume of the Mississippi River. *J. Expt. Mar. Biol. Ecol.* **126**, 21–43.

Vanderploeg H.A. (1990) Feeding mechanisms and their relation to particle selection and feeding in suspension-feeding zooplankton. In *The Biology of Particles in Aquatic Systems* (ed. by R. Wotton), pp. 183–212. CRC Press, Boca Raton, FL.

Vanderploeg H.A. & Paffenhöfer G.-A. (1985) Modes of algal capture by the freshwater copepod *Diaptomus sicilis* and their relation to food-size selection. *Limnol. Oceanogr.* **30**, 871–5.

Vanderploeg H.A., Paffenhöfer G.-A. & Liebig J.R. (1988) *Diaptomus* vs. net phytoplankton: Effects of algal size and morphology on selectivity of a behaviorally flexible, omnivorous copepod. *Bull. Mar. Sci.* **43**, 377–94.

Vanderploeg H.A., Paffenhöfer G.-A. & Liebig J.R. (1990) New cinematographic observations and hypotheses on the concentration-variable selectivity of calanoid copepods for particles of different food quality. In *Behavioural Mechanisms of Food Selection,* (ed. by R.N. Hughes), *NATO ASI series, vol. G 20,* pp. 595–614. Springer Verlag, Berlin.

Werner E.E. & Hall D.J. (1974) Optimal foraging and the size selection of prey by the bluegill sunfish *Lepomis macrochirus. Ecology* **55**, 1042–52.

7: Gourmands of Mud: Diet Selection in Marine Deposit Feeders

PETER A. JUMARS

INTRODUCTION

Deposit feeding, a subset of detritivory, is the eating of sand and mud. The central question in deposit-feeding research remains largely unanswered: What biological and chemical components of sediments are assimilated by deposit feeders? Candidates in specific settings and in spatially and temporally varying combinations include bacteria and their exudates, protozoa, (both benthic and sedimenting planktonic) microalgae, non-living particulate detritus, and interstitial solutes. Examinations of shallow-water sediments suggest that bacteria cannot constitute the principal carbon source for deposit feeders (Cammen 1989) but may be important or even dominant in reduced nitrogen supply. To date it has been established that bacteria are digested efficiently in many deposit feeders, that a few intertidal deposit feeders specialize on digesting and assimilating the contents of microalgae, and that at least a few deposit feeders can absorb organic matter from non-living, organic detritus (Lopez *et al.* 1989). Nitrogen rather than carbon limitation of growth rate seems more likely in this detritus-based system, so nitrogen would appear to warrant more focus than carbon. The inherent, intertwined problems in identifying the resources used by deposit feeders are that at their prodigious feeding rates a volumetrically and gravimetrically very small component of ingested sediments could yield the bulk of assimilated energy or matter (Cammen 1989) and that sediments are physically, biologically and chemically heterogeneous. Despite the fact that precise, accurate and general answers to the central question are still elusive, the range of possibilities for the unstudied majority of deposit feeders has been narrowed considerably by growing knowledge of sedimentary organic chemical characteristics and kinetics of deposit feeding.

The purpose of this chapter is to examine existing data on selection by deposit feeders in the context of both these constraints and any

124

constraints that can be imposed from general foraging theory. The time constraints imposed by large volumetric and gravimetric throughputs seem very different than for the macrophagous vertebrates treated by early foraging theory (Stephens & Krebs 1986). An analogy suggested by this examination of possibilities and constraints is of a rapidly operating machine that balances largely fixed and mechanical selective abilities with very flexible volumetric processing rates to make a nutritional profit under the majority of natural circumstances. This analogy also raises the pointed question of whether naive predictions of what should be preferred for ingestion fall within the capabilities of the machine. Testing of the machine analogy provides perhaps the most fundamental reason for study of deposit feeders; they appear to represent an extreme foraging strategy in rate of processing of food, fraction of their time devoted to food processing and importance of digestive and absorptive kinetics to their fitness. The marginal value and its variant, the principle of lost opportunity (Stephens & Krebs 1986), prove especially useful for understanding the limits on selection set by the need for a fast rate of throughput. Some gestalt for this evolutionary focus on feeding can be gained from the epithet 'roving gut' that has been applied to deposit feeders (especially the worms) by vertebrate zoologists. Indeed, up to 80% of apparent body volume can be taken up by the gut.

DEFINITIONS AND SUBDIVISIONS OF THE GUILD

Since assimilated food has not been identified with certainty for many deposit feeders, the definition of deposit feeding must be based instead on characteristics of ingested material. The most commonly accepted definition includes those animals that **frequently** ingest sedimented material of low **bulk** food value (Jumars *et al.* 1984; Lopez & Levinton 1987). The terms in bold are exceedingly important in applying the definition, but use of these discriminant variables remains very subjective for want of quantitative data. Few predators living in sediments can avoid ingesting sediments incidentally to prey capture, yet they clearly cannot be considered deposit feeders. Likewise, few meiofauna (animals retained on a 40-µm sieve but passing a 300–1000-µm sieve) that specialize on ingesting bacteria, microalgae or protozoa can avoid incidental ingestion of sediments. For kinetic reasons that will be elaborated below, however, it is doubtful whether meiofauna – or any animals with gut volumes much smaller than about 0.1 mm^3 – can be deposit feeders.

Two subcategories of deposit feeders are recognized frequently on the basis of the sedimentary horizon from which particles are ingested. Surface deposit feeders feed at the sediment–water interface, while subsurface deposit feeders feed below it. Although this definition seems clear enough, there is an intermediate category of deposit feeders, called funnel feeders, that feed with their anterior ends below the mean position of the sediment water–interface and thereby cause sufficient slumping of material downward to create, at least on occasion, funnel-shaped depressions in the sediment–water interface. An advantage to subsurface and funnel feeders in intertidal settings is that they can continue to feed after the tide is out if capillary water remains. Surface deposit feeders can do so only if overlying water remains or else must feed from void (e.g. burrow) surfaces below the plane of the sediment surface. If funnel feeders eat fast enough, they get mostly surficial deposits, with deposition of some kinds of particles from suspension enhanced by the presence of the pit itself (Nowell *et al.* 1984; Yager *et al.* 1993). A further potential problem with the surface–subsurface dichotomy is that animals may feed on surfaces, such as burrow walls, below the sediment–water interface with the aid of appendages much like those that are seen in animals that feed at the sediment–water interface. Pectinariid polychaetes (ice-cream-cone worms), protobranch bivalves and even some terebellid polychaetes (Nowell *et al.* 1989), for example, feed with tentacles below the sediment–water interface. Thus the morphological characters normally used to assign feeding guilds (e.g. Fauchald & Jumars 1979) may not be reliable indicators of feeding stratum.

Motility categories are also sometimes recognized (e.g. Fauchald & Jumars 1979). There are so few data, however, on the movements of individual deposit feeders that such classifications can be neither very detailed nor very accurate.

THE CHEMICAL, PHYSICAL AND GEOLOGICAL ENVIRONMENT AS A DETERMINANT OF FOOD QUALITY AND QUANTITY

Although the character of absorbed food remains poorly identified for marine deposit feeders in general, the form of organic matter for detritivores is certainly very different in terrestrial vs. open-ocean marine settings. Labile, nitrogen-rich plant protoplasm rarely arrives in the terrestrial litter community; plants resorb much of these valuable components before leaf abscission. In the marine realm, however, whole microalgal cells seasonally or routinely do arrive at the sediment–water surface, even in deep water (Billett *et al.* 1983). The most labile material for open-ocean deposit feeders thus is newly arriving.

Cellulose (with lignin) clearly dominates terrestrial inputs of organic matter. Foregut fermenters (e.g. ruminants) and hindgut fermenters (e.g. termites) have evolved morphologically obvious means to tackle structural carbohydrates directly, but most litter organisms have not. Thus while available calories per mole of carbon must eventually decline with time after food inputs in either the marine or terrestrial realm, food value per gram of terrestrial organic matter for most litter organisms probably has peak value at some intermediate time after litter fall when microbes have degraded the polymers and have added at least their own masses of labile nitrogen. With the caveat that a few important exceptions (i.e. cellulose, chitin and structural carbohydrates from macroalgae) may be more useful after some ageing (Tenore & Hanson 1980) and may foster fermentative associations (e.g. Fong & Mann 1980), the bulk of organic inputs to marine sediments would appear most valuable to deposit feeders immediately upon input. Microalgae have much less mass in structural, polymeric carbohydrates than do terrestrial plants. Geochemical studies support this argument by documenting that a substantial part of the bottom-arriving particulate organic flux degrades quickly (Reimers 1989). This observation suggests strong natural selection for using or sequestering labile organic matter as soon as possible after its arrival.

Two major and strongly interacting physical differences between terrestrial litter communities and marine deposit-feeding communities involve excess density and the capabilities of the respective fluids to move both particles and solutes. The force required to lift a particle is linearly proportional to its excess density, i.e. its particulate density minus that of the fluid in which it is immersed. Moving water, by virtue of both its greater inertia and greater viscosity, is far more effective at transporting and redistributing particles of even the same excess density than is moving air. It is easy to forget when viewing sediments exposed at low tide that immersed sedimentary environments are scenes of constant, periodic or episodic particle motion. Newly arriving organic material at the seabed has lower excess density on average than the mineral grains that compose the gravimetric bulk of the deposit and thus is more easily redistributed. Density separations with surficial sediments immersed in high-density fluids reveal that the low-density fraction (<1.9 g cm^{-3}) is enriched by a factor of 10 to 100 in organic content per unit of weight (L. Mayer *et al.* in review). Mayer (1989) has suggested that ageing of organic matter at and in the seabed corresponds, with due accounting for the material lost to mineralization, with a transition from labile organic particles to refractory, monomolecular coatings on mineral grains.

There is a long- and well-established relationship between grain surface area and several other variables important in determining food quality for deposit feeders: organic content (Longbottom 1970), microbial metabolic rate (Hargrave & Phillips 1977) and microbial abundance (Dale 1974). DeFlaun and Mayer (1983) have refined the latter relationship by pointing out that bacteria tend not to attach to particles smaller than about 5 μm; while their study is limited to one intertidal location, it seems logical that the value to bacteria of attachment in general would fall as the size of the particle approaches their own cell size. This surface-area relationship implies that food value per unit of volume of food ingested scales roughly as grain diameter $^{-1}$, at least down to the grain size at which microbial attachment per unit of grain area declines.

Implicit in this discussion is that the sediments in question are non-cohesive. Watling (1988) points out, however, that muddy sediments are much more like a complex sponge or lumpy gel than they are a collection of easily separable grains. An unsolved problem, then, is the extent to which sediments behave like and are perceived as (by deposit feeders) individual grains rather than as aggregates of grains. In the latter case, it is the properties of aggregates and not individual grains that limit selectivity. In general, it can be expected that cohesion and adhesion among grains will decrease the ability of animals to select among particles that constitute such aggregates.

Physical, chemical and biological components of the environment interact to set limits on the rate of supply to, and thus the potential for selection among particles by, deposit feeders (Fig. 7.1). At one extreme of possibilities is the now classic view of marine detritivory developed by Newell (1965) and extended by Levinton & Lopez (1977) based on the observation that fecal pellets after ageing and disaggregation yield particles worth eating again. In this situation fluxes in and out of the individual's ambit are ignored (i.e. implicitly assumed to be minor or of equal magnitude and opposite sign); this closed-system view does accurately portray the laboratory experiments on which these studies have largely been based. Food abundance (number of particles) is set primarily by the ratio of the rates at which sediments are ingested to the rate at which pellets disaggregate (Levinton & Lopez 1977), while food quality is set by the extent of microbial addition of labile organic nitrogen to the

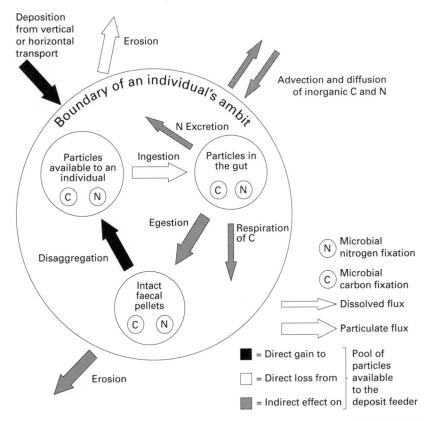

Fig. 7.1 The ability of water to transport particles and solutes makes food availability within the ambit of an individual marine deposit feeder far more dynamic than one might expect from experience with a terrestrial litter community. Not only are erosion and deposition frequent, but microbial (including microalgal) nitrogen and carbon fixation are also influenced by fluid transport.

disaggregating pellets (Newell 1965). This extreme falls close to what can be expected of terrestrial litter communities, where the time scale of particle inputs greatly exceeds that of fecal pellet breakdown. To provide much choice among particles at a steady state, the disaggregation rate must exceed the ingestion rate.

At the other extreme, physical transport constantly delivers or exchanges particles, swamping ingestion rate in magnitude and providing substantial food supply rate and potential choice to the deposit feeder. Because of the ability of combined waves and currents to move sediments on scales approximating those of individual ambits, intertidal communities in which sediment transport measurements have been carried out (Grant 1983; Miller & Sternberg 1988) fall much closer to this extreme. There is reason, then, to expect diet selection by at least some deposit feeders. This perspective also raises a critical difficulty in evaluating selectivity by surface deposit feeders; at one extreme they may simply use the ambient sediments as a residence and feed on the flux of material going by, making comparisons of ambient sediments and diets of little relevance to the issue of diet choice. In this scenario, the flux (a rate) of particles may also determine feeding rate (Brandon & Miller, in preparation).

STATICS, KINETICS AND SIZE SCALING OF DEPOSIT FEEDING

Although studies of item and patch choice dominated early research on diet selection by deposit feeders as they did with macrophages, feeding rate so pervades these issues that considerable space is saved by treating this often overriding constraint first. In treating ingestion rate, it is tempting to create an implicit analogy with large grazers (Chapter 3). Herbivores on poor forage are well known for the prodigious rates at which they process food (Van Soest 1982), but they pale by comparison with deposit feeders. Deposit feeders typically ingest three times their own (dry) weight in (dry weight of) sediments per day (Fig. 7.2); even with allowance for the high bulk density of sediments, that figure translates to a volumetric rate in excess of a full body volume per day. Their maximal gravimetric rates of 10^2 body weights per day (Fig. 7.2) correspond with animals that ingest food particularly dilute in organic content.

Some of the high gravimetric and volumetric ingestion rate is achieved via a gut volume that exceeds in proportion of total body volume that of closely related animals (Fig. 7.3, *see* page 132), but the extreme volumetric and gravimetric throughput rates of deposit feeders are achieved primarily via short gut residence times. There is no strong correlation among species (Fig. 7.3) between body size and proportion of the body occupied by the gut. Similarly, with the important proviso that small juveniles were not sampled, Penry & Jumars (1990) found within deposit-feeding species that the proportion of the body occupied by the gut generally changes isometrically with body size. Figures 7.2 and 7.3 combine to provide a strong contrast with ruminants. Larger ruminants have greater fractions of their metabolic weights (weight$^{0.75}$) devoted to digestion (Table 1 in Hoppe 1977). Poorer forage characteristic of large

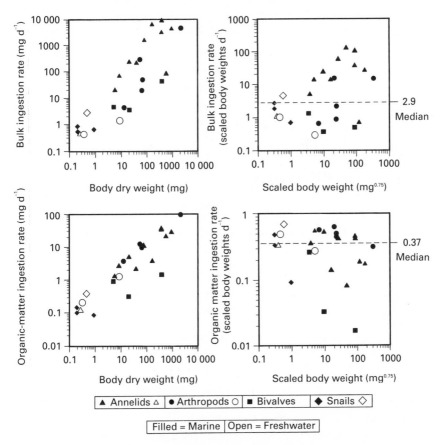

Fig. 7.2 Bulk ingestion rate of sediments and organic-matter ingestion rates (both as dry weights) versus body dry weights of deposit feeders, plotted from the data tabled by Cammen (1980, 1987). He was careful to select data from animals observed near 15°C, with each datum representing an average for individuals of one species and one average size. In the panels on the right the value of (body weight)$^{0.75}$ rather than unmodified body weight is plotted to remove the expected trend with body size and to reveal that the extremes of rapid ingestion rate correspond with food poor in organic matter.

grazers takes longer to ferment, and this residence-time constraint (Dement & Longhurst 1987) necessitates the greater volume (Chapter 8). By contrast, poorer food for deposit feeders drives faster feeding when species (not individuals) are compared over a large range in food quality (Cammen 1980).

When closely related deposit feeders are compared, there is some evidence of greater gut volume in animals living on poorer food, and isometry of gut and body volume coupled with the usual scaling of

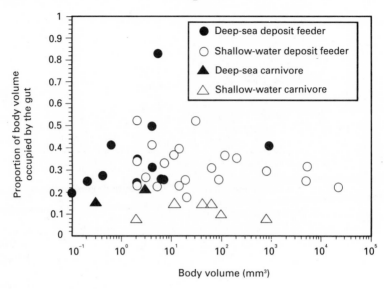

Fig. 7.3 Proportion of the body volume occupied by the gut versus body volume for a range of marine polychaetes, drawn from the data tabled by Penry and Jumars (1990). Deposit-feeding polychaetes typically have about ⅓ of the body volume occupied by the gut, but ½ is not unusual, and one species, the cirratulid *Tharyx luticastellus*, devotes over 80% of its body volume to the gut. There is no dramatic correlation of body volume with proportion occupied by the gut.

ingestion rate with body weight (weight$^{0.75}$) does imply somewhat longer gut residence times in larger individuals and species. Feeding rates of the larger deposit feeders (e.g. the larger holothuroids) are not well studied enough, however, to allow their placement on Fig. 7.2. Characteristic gut residence times of deposit feeders in the range of sizes represented in Figs 7.2 and 7.3 are 0.5–6 h. To bring the fermentation contrast down to more comparable body size, a termite that feeds on **refractory but ultimately digestible** lignocellulose spends up to 48 h processing before releasing the particulate residue (Bignell 1984). Paradoxically, the short residence times of material in the guts of deposit feeders must make these animals specialists on digesting and absorbing relatively **labile** organic components of their food. The contrast is between a dilute and labile food resource in deposit feeders and a concentrated but refractory resource in fermenters that repays the costs of mutualism (Plante *et al.* 1990). This conclusion suggests that replacing the ordinate of the lower right panel of Fig. 7.2 with a measure of labile rather than total organic matter would both lower the median amount ingested per day and remove much of the scatter.

Scaling of nutrition in early life is a widespread problem. It is acute in deposit feeders because of the switch from relatively rich larval resources to the exceptionally dilute ones that deposit feeders utilize. Forbes and Lopez (1990) documented such a break in allometry for the polychaete *Capitella* sp. I. Cammen (1980, 1987) found no deposit feeder of <1 mg dry weight ingesting <13% organic matter. Gallagher *et al.* (1990) found strong circumstantial evidence that juveniles of the deposit-feeding (as an adult) polychaete *Hobsonia florida* specialize on diatoms. Individuals of the fiddler crab *Uca longisignalis* show a switch from discrete zooplankton prey as zoea to more dilute diatoms of sediments as adults (Weissburg & Zimmer-Faust 1991). The insoluble problem at small body size if the food is dilute is that there is not sufficient gut volume to provide a substantial throughput rate (volume per unit of time) with sufficient residence time to allow adequate digestion and absorption.

Although consensus has not been reached on the controls governing feeding rates in even adult deposit feeders, numerous experiments reveal great variability among individuals in feeding rate on identical foods as well as great flexibility in feeding rate of individuals from diet to diet. It is not known whether the variability among individuals is due to variability in prior diet vs. diversity in genetic composition. Whatever its source, this variability constitutes part of the scatter in the relationships of Fig. 7.2, but flexibility on varied diets is expressly avoided in the choice of data shown here, and among-individual variation is limited by using only means for each species. One way of looking at the trend in Fig. 7.2 is that it summarizes the mean rate of feeding in deposit feeders of a given size on the sediment to which they have adapted over evolutionary time. The inverse relationship between feeding rate and organic content of ingested sediments over several decades in both variables is a reflection of the absence of a free lunch; to support an average amount of organic matter (body size) takes an average rate of throughput. Since the plot is of means for given species and thus represents among-species patterns, however, it bears little direct relevance to the issue of the rate at which an individual does or should feed when food value is changed from the mean of the environment to which it is adapted. Because species adapted to sediments of lower organic matter must in the mean feed faster to make ends meet says little about what they should do when presented with an enriched food resource.

Taghon and Greene (1990) presented a set of data for one lugworm species (*Abarenicola pacifica*) and a detailed analysis of fit to the broad suite of models that have been proposed to explain or predict individual feeding rates as a function of the quality of food presented to an individual. They analyzed individual feeding and growth rates on sediments of varying protein contents and two very different protein sources, finding maximal feeding rate at a sediment protein concentration of approximately 0.1 mg protein (g sediment)$^{-1}$. Growth rates, on the other hand, showed monotonic increase with increasing protein concentration. It is extremely important that this unique data set, combining data on individual growth and feeding rates, be replicated with additional species. Among extant predictive models for deposit feeders (i.e. Taghon 1981; Phillips 1984; Penry & Jumars 1987; Kofoed *et al.* 1989; Dade *et al.* 1990), their results are consistent with only the formulation of Dade *et al.* (1990), who coupled Michaelis–Menten kinetics of digestion with Michaelis–Menten kinetics of absorption, adopting the premise that animals adjust ingestion rate to maximize the rate of absorption. For any anticipated digestive kinetics and absorptive kinetics at one set food concentration (moles of food per unit of weight of sediments) and unlimited food quantity (weight of sediments) there is an optimal throughput time. Longer residence times produce lower rates of absorption because they result in slowed rates of digestive reaction (Penry & Jumars 1987), while shorter ones drive rapidly produced products out of the gut before they can be absorbed. For the complex (higher-order Michaelis–Menten in the terminology of Dade *et al.* 1990) digestive kinetics of deposit feeders, these optimal ingestion rates should be slow at low food concentrations because it takes time to produce enough digestive products to drive absorption, should rise until the absorptive system is saturated, and then should fall again as longer throughput times suffice to maintain absorption rate at its maximum (Dade *et al.* 1990). Net absorption rate, however, should increase with increasing concentrations, as suggested by the growth rates seen by Taghon and Greene (1990). To avoid confusion in the face of terminology that varies considerably among authors it is worth underscoring that concentration of food here is taken as a measure of its quality and that the model predictions assume that food of the indicated quality is available in unlimited amount.

Earlier thinking that failed both to distinguish and to couple digestion and absorption suggested (Taghon 1981; Penry & Jumars 1987) that faster ingestion rate would always produce a higher gross rate of gain because Michaelis–Menten digestive production rates decrease monotonically with gut residence time. The idea and problem are still typical of optimal foraging approaches. The apparent rate of gain (digestive production) could be formulated relatively easily, but poorly specified or unknown and non-linearly increasing costs of increased feeding rate had to be invoked to avoid the absurd prediction of infinitely fast feeding. Two possibilities that seemed likely in view of the prodigious feeding rates of deposit feeders were the costs of mechanical processing (weight of sediments moved a given distance and height per unit of time) and costs of digestive enzyme production. Taghon (1988) devised a means to partition these two variables and anabolic costs (specific dynamic action) in a system of three equations in three unknowns. His surprising experimental result, again with the lugworm *Abarenicola pacifica*, was that even for individuals feeding on sediments of no food value the mechanical and enzymatic costs of feeding were insignificant. Taghon's (1988) and Taghon and Greene's (1990) results appear to support the idea that deposit feeders, for which quality rather than quantity of food seems to be the principal limitation on growth rate, are free to set their throughput rates at the absorption-rate maximum suggested by Dade *et al.* (1990). If digestive costs or mechanical costs of feeding were large, then this absorption-rate maximum might not correspond with maximal rate of energy or mass gain. A corollary (Dade *et al.* 1990) of feeding at a rate set by absorption is that gut residence time becomes a far more important determinant of diet choice than does pre-ingestive handling time.

Ingestion rate of deposit feeders clearly is flexible (e.g. Taghon & Jumars 1984), but it is not yet clear what signals the animals use to set this rate. Dade *et al.* (1990), to derive their model, assumed that deposit feeders maintain a full gut but respond with ingestion (and egestion) rate changes to their rate of absorption of digestive products. It seems very likely, however, that at least four sets of stimuli operate to affect ingestion rate, since they do in many other animals from insects (Bernays 1985) to ungulates (Illius & Gordon 1990). These stimuli are, in sequence, smell, taste, distension of the gut and internal detection of the levels of absorbed products in body fluids. Here smell is defined operationally as chemical detection of food prior to its contact by an appendage. Although working with intertidal organisms during emersion precludes ready separation of smell from taste and the 3-h duration of experiments includes time for feedback from absorbed food, the most complete data on chemical stimulants of feeding rate are for the sand fiddler crab *Uca pugilator* (Robertson *et al.* 1981). L-serine, sucrose and maltose proved highly stimulatory, consistent with the crab's dietary specialization on benthic diatoms. Qualitative experiments easily reveal that smell does affect ingestion rate of at least some deposit feeders. In experiments with selection on valueless particles (glass beads), for example, in Jumars' laboratory seawater extracts from complex food mixtures (the commercial aquarium preparation Tetramin$^{®}$) are routinely applied to elicit adequate ingestion rates. The normal procedure (e.g. Self & Jumars 1988) does not distinguish smell from taste, but it has been observed (unpublished observations) that the dissolved form alone when dispensed near the polychaete *Pseudopolydora kempi japonica* immediately stimulates feeding palp activity.

There are phagodepressants as well. Valiella *et al.* (1979) documented that cinnamic acids at natural concentrations inhibit detritivore ingestion. The experimental protocol again made separation of smell from taste impractical. A common observation of laboratory and field feeding traces and bottom photographs is that feeding marks (e.g. grooves from deployment of tentacles) avoid fecal pellets of the same species. Forbes and Lopez (1986) found that recently egested, disaggregated fecal material depressed feeding rates in the snail *Hydrobia truncata*. Miller and Jumars (1986) similarly showed that accumulation of pellets of *Pseudopolydora kempi japonica* in its feeding area slowed its feeding rates. It may be more practical from the standpoint of sensory capability for a detritivore able to gain nutrition from a diversity of foods to recognize lack of resupply (accumulation of fecal pellets) rather than carrying all the detectors necessary to recognize resupply. Perhaps the most bizarre situation, however, has been seen in the subtidal and deep-sea polychaete *Amphicteis scaphobranchiata*. Nowell *et al.* (1984) documented that it literally slings its fecal pellets outside its feeding zone, thereby creating a feeding pit that traps material in suspended and bedload transport. In the slinging activity the worm is at great risk to predation, since two-thirds of its body is extended from the protective tube. A reasonable scenario for the evolution of this behaviour is that individuals that ejected pellets from their feeding area both removed a phagodepressant and enhanced food supply by making a pit, contributing sufficiently to rate of nutrient gain to outweigh the predation risk.

Feeding appendages of surface deposit feeders constitute major components of the diets of many bottom-dwelling fishes, and thus predation also modulates feeding rate by affecting time spent feeding. Levinton (1971), for example, established that *Macoma tenta*, a tellinid bivalve, was active only at night when visual predators were less effective. Such periodicity may not be limited to surface deposit feeders alone, since many subsurface feeders place the opposite ends of their bodies at risk when they come to the sediment surface to defecate. Fuller *et al.* (1988), for example, report greater nocturnal feeding rate in the capitellid polychaete *Mediomastus ambiseta.* Some deposit feeders reduce the risk of predation and thus ameliorate the time constraint by incorporating toxins in their feeding appendages and other structures that often protrude from the tube (e.g. Gibbs *et al.* 1981).

Another modulator of feeding rate is resupply of high-quality particles. Nichols *et al.* (1989) found that the heart urchin *Brisaster latifrons,* that normally burrows several centimetres below the sediment surface, emerged in apparent response to modulation by current reversals of the supply of rich surficial particles. Jumars and Self (1986) documented enhanced feeding rate in one species of surface deposit feeder just after a sediment transport event but found no evidence of a similar effect in a sympatric species. Both of the tentacle feeders studied by Jumars and Self (1986) are constrained mechanically to feed only when immersed, but a substantial fraction of the flat is covered by shallow pools at low tide. Conversely, intertidal ocypodid crabs, like the fiddler and sand bubbler crabs discussed below, are constrained to feed at low tide. Penry and Jumars (1987) have suggested that animals constrained to feed less frequently should choose for digestion items that continue to give gain until the animals are able to feed again.

PATCH EXPLOITATION

Feeding rate thus should in part be determined by the ability of deposit feeders to locate or produce (by affecting sediment transport or microbial growth) patches of high food quality. Most deposit feeders, however, are quite restricted in mobility compared with cruising predators and scavengers. A major component of habitat choice is therefore associated with larval or juvenile settlement at a site. Deposit feeders have long been known to recognize and prefer bacteria-coated sediments to cleaned ones (Meadows 1964), a preference that makes good sense with respect to food value. A more recent suggestion that has stimulated considerable debate is that a major component of site selection may be passive (Butman 1987). To overstate the case, a deposit feeder might find its way to a good feeding environment by mimicking in sediment dynamic properties (i.e. settling velocity) the characteristics of its preferred food. While this null model is an interesting one, departures from it already are known (Butman *et al.* 1988), and more can be expected. One problem is that the bulk sediments represent responses primarily to extreme conditions that may exist only episodically during spring tides or major seasonal storms, and it is not clear, therefore, that larvae responding to physical conditions at one time would end up reliably in the proper regime. A further problem is that food for recently settled juveniles may not match food for adults, but resuspension and resettlement of juveniles may alleviate both these problems (Emerson & Grant 1991). At the very least, however, the null model has injected some physical realism and controversy that is spurring research on the issue of larval habitat choice. Analogy with ballooning gives a vivid impression of the difficulty of habitat choice. Control is primarily in vertical position in the winds or currents and is limited by the fuel onboard or energetic reserves. Thus, the decision is not where to go, but when to stop. Marine larvae are well known to become less selective of settlement sites as time proceeds. The problem is eminently suited to dynamic programming as a theoretical approach (Mangel & Clark 1988), but has not yet been formulated in the context of a choice that affects subsequent foraging success.

As one of the two initial problems addressed by (optimal) foraging theory, it is not surprising that more information is available with respect to the more classically defined patch choice. Attention among students of trace fossils to rules for space utilization in fact predates foraging theory by nearly forty years. Richter (1928, cited in Raup & Seilacher 1969) observed that many fossil traces seemed to conform to a simple, three-part algorithm: turn approximately 180° after going approximately a given distance, avoid recrossing (i.e. avoid previous tracks), and keep close proximity to previous tracks. More recent computer simulations of trace fossils have shown that slight changes in these very simple rules can generate diverse patterns (Papentin 1973). More importantly for the present context, some of the most frequently observed patterns (e.g. Fig. 7.4a) of burrowing or crawling traces are consistent with the marginal value theorem, i.e. that an individual should forage in a patch only until its mean rate of gain falls to the mean level that can be expected from the environment as a whole (including the search time needed to find a new patch). Another suite of observations consistent with foraging theory of patch choice is those of Scheibling (1981) concerning a microphagous seastar. It apparently uses flow direction as a cue to avoid recrossing already foraged territory. An explicit, experimental test of patch choice for deposit feeders was that of Robertson *et al.* (1980) on the fiddler crab *Uca pugnax*. They indeed found that crabs spent more time foraging in patches that had been manipulated to have higher concentrations of benthic diatoms and by probing with sensory setae on their legs could

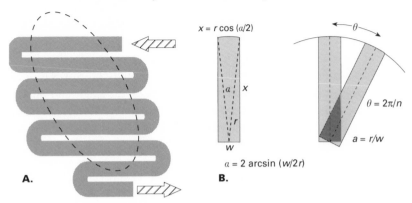

$$x = r \cos (a/2)$$

$$\theta = 2\pi/n$$

$$a = r/w$$

$$a = 2 \arcsin (w/2r)$$

A.   **B.**

Fig. 7.4 Two variants of Richter's (1928, cited in Raup & Seilacher 1969) rules for patch utilization by deposit feeders. **(A)** A hypothetical trace of a mobile deposit feeder encountering a food-rich patch (dashed ellipse). It proceeds straight forward until a food-rich patch is encountered and doubles back whenever it leaves a rich patch. **(B)** Definition sketch for Ohta's (1984) model of patch utilization by an echiuran worm feeding by making radial strokes of a given aspect ratio, a, from a central burrow. Strokes are spaced evenly, i.e. $2\pi/n$ radians apart where n is the number of strokes. Note that overlap (dark grey shading) is extensive near the centre of the circle for large θ, while overlap of strokes is minimal at the distal portions of strokes. For simplicity, Equation 7.2 calculates the unharvested (white) area between adjacent strokes.

resolve patches of millimetre scale. Forbes and Lopez (1986) found similarly in the snail *Hydrobia truncata* that animals spent more time foraging in patches richer in chlorophyll. Snails slowed their rates of crawling when they encountered higher chlorophyll concentrations but did not change egestion rates significantly. An elegant analysis of patch utilization (Weissburg, in press) also has recently been completed that provides strong support for the existence of patch-leaving resource thresholds in deposit feeders.

These various patterns and experiments involve mobile animals and thus require no fundamental changes to the notions of patch choice for non-deposit feeders. In a very innovative contribution, Ohta (1984) added the geometric constraints of foraging with a single tentacle from a fixed burrow location to the idea of patch choice for deposit feeders. He analysed the number of strokes made by the single feeding proboscis of deep-sea echiurans (gutter worms) evident in bottom photographs and observed a remarkably strong relationship ($r^2 = 0.95$ for a regression based on 14 points) between the maximal number of strokes seen (n_m) and the aspect ratio, a (length/width or radius/width, r/w), of the strokes and presumably of the appendage that produces them:

$$n_m = 4.86 + 2.39a \tag{7.1}$$

For the two extremes of a seen, i.e. 4 and 14, this equation gives $n_m = 14.42$ and 38.32, respectively. While the geometry of strokes is simple conceptually, it is messy computationally because of the single and multiple overlaps near the centre of the circle. These overlaps make it easier to calculate the fraction, f_u, of the circle that is left uncovered by n evenly spaced strokes than it is to calculate the fraction of area with overlaps. Specifically,

$$f_u = 1 - \frac{n}{\pi}\left[\arcsin\frac{1}{2a} + \left[\frac{\sin^2\left(\frac{\pi}{n} - \arcsin\frac{1}{2a}\right)}{\tan\frac{\pi}{n}}\right] - \left[\frac{\sin^2\left(\frac{\pi}{n} - \arcsin\frac{1}{2a}\right)}{\tan\left(\frac{\pi}{n} - \arcsin\frac{1}{2a}\right)}\right]\right] \tag{7.2}$$

Substituting for n the values of n_m calculated for a = 4 and 14, respectively, gives f_u = 0.18 and 0.32 (not a constant value of 0.24 as suggested by Ohta). The number of strokes, n_c, that it would take to just cover the feeding circle completely (distal edges of strokes just touching) is calculated far more easily by noting that for such, just complete, cover $a = 0$ of Fig. 7.4b. It also corresponds with the number of chords of length w that would be needed to go fully around the circle:

$$n_c = \frac{\pi}{\arcsin\frac{1}{2a}} \tag{7.3}$$

Complete overlap (each spot touched by at least two strokes) would not occur until the number of strokes reached $2n_c$. Because of the uniform geometry of the problem (covering a circle with radially arranged rectangles), 75% cover of the circle by strokes (f_u = 0.25) corresponds approximately with $n_c/2$ for the full range of a documented by Ohta (1984). The generalization is approximate because the real number of strokes must be an integer, while Equations 7.1 to 7.3 in the simple forms written allow fractions. (The approximation gets far worse for a < 0.71.) There is rough correspondence between n_m of Equation 7.1 and $n_c/2$ over the full range of a seen, but for high aspect ratios 75% cover is reached at higher numbers of strokes than are observed (Fig. 7.5); $n_c/2$ overestimates the number of strokes observed (Equation 7.1) by about 15% at a = 14 and equals n_m only for a = 6.49. While $n_c/2$ and n_m both are linearly related with a over the domain of interest, $n_c/2$ has a somewhat steeper slope.

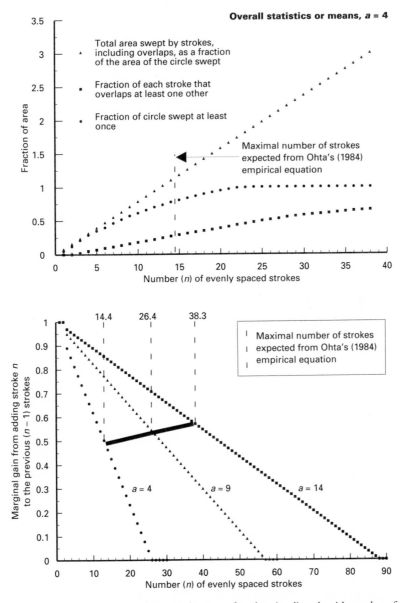

Fig. 7.5 An example for $a = 4$ of the way that area of strokes rises linearly with number of strokes while fraction of the circle harvested (touched at least once) and average fraction of a stroke that consists of area overlapping another stroke both rise more slowly. Complete coverage of the circle is achieved at the number of strokes given by Equation 7.3, while complete overlap does not occur until twice that number of strokes. Evolution probably acts upon the marginal gain achieved by adding another stroke rather than upon the mean overlap per stroke. Shown are such marginal gains for the full range of aspect ratios (4–14) reported by Ohta (1984). The animal appears to stop adding strokes when about 50% of the added stroke constitutes previously unharvested sediments (heavy line).

The tight fit of Equation 7.1 as well as this departure from strict geometric similarity with changing aspect ratio still begs for evolutionary explanation. It is not clear that the animal has any way to evaluate mean quantities for the circle as a whole. It seems more likely that evolution has operated upon the marginal gain per stroke (Fig. 7.5) than upon integral measures for the circle, i.e. upon the difference in harvested area (h), expressed as the fraction of an additional stroke's area that would contact previously unharvested sediments with increasing number of strokes (n) or $\Delta h/\Delta n$. This function has no peak and declines linearly with n (Fig. 7.5). The previously untouched area contacted, when expressed as a fraction of the area of one stroke, drops to near 50% for $n_c/2$ and thus also for low values of a in Equation 7.1, but stays higher at about 60% for n_m of Equation 7.1 when $a = 14$ (Fig. 7.5). The disparity is probably due to imprecision by the animal in locating strokes exactly θ radians apart, this imprecision compounding when n is large. The likelihood of some imprecision makes the estimates of harvested area and marginal gain under the assumption of perfectly evenly spaced strokes upper bounds and the estimates of overlap lower bounds, at least until full cover is expected (Equation 7.3). If there is a given variance associated with placing each additional stroke, then one would expect overall imprecision to grow approximately linearly with n, returning an increasingly lower marginal gain than predicted as n increases. Thus, the marginal gain for values of n_m given by Equation 7.1 and true values of f_u may be relatively constant if this imprecision of stroke placement is included in the problem.

There is reason to expect echiurans and sipunculans (peanut worms) to be particularly sensitive to this marginal gain. Jumars *et al.* (1990) and Plante *et al.* (1990) have argued that echiurans and sipunculans are likely to cache recently arrived deposits for later use. The cache may be in the form of the animal's own, microbially modified, fecal pellets. Although caching gives access for nutritional purposes to external volume in addition to gut volume or other somatic storage and thus also relieves some volumetric and temporal constraints upon digestion, cache volume must be limited and should not be filled with 'old' material of low bulk food value. The rule of spacing strokes at equal angles greater than the one that would yield complete coverage of the circle is an example of a simple way to achieve recovery of a reasonably large proportion of newly deposited material without recourse to sensory evaluation of particles one by one. The need for 'new' material also raises questions of risk sensitivity of the gathering strategy, since bloom sedimentation events on the deep-sea floor would appear to be less than exactly predictable (Billett *et al.* 1983).

Further, fascinating problems include the prediction of both aspect ratios (a) and absolute reaches (r) of feeding appendages. One might expect both to increase with decreasing flux of food since the radius is the reach and a greater aspect ratio gives a greater reach from the same material investment (area and presumably mass of tentacle). Imprecision in tentacle placement may set the upper bound on a for species that rely on contacting unharvested surface, but non-caching species, e.g. of spionid and terebellid polychaetes, have much larger values of a than 14 (and from 2 to 10^2 feeding palps or tentacles). High a is associated with multiple tentacles, presumably so that a reasonable fraction of the feeding area can be explored and harvested per unit of time. The lower bound on aspect ratio, on the other hand, probably stems from the limited gain in reach achieved and from acute overlap problems when $a < 0.71$. Nor should even coverage of the circle be considered always ideal. In a classroom experiment J. Grebmeier (personal communication) found the tentacle-feeding spionid *Pseudopolydora kempi japonica* to spend disproportionate time feeding in food-enriched sectors of its feeding circle. Such patch-selective capability is consonant with flow-induced heterogeneity in food quality (Eckman & Nowell 1984).

Perhaps the best example of strong constraints acting on patch exploitation, however, is in foraging by an intertidal crab that belongs to the same family as fiddler crabs but, unlike them, apparently foregoes evaluation of patches. Zimmer-Faust (1987, 1989) documented that the sand bubbler crab, *Scopimera inflata*, does not select patches of high food value. It lives in an intertidal sand zone where food concentrations are low and food presumably is redistributed by wind-wave induced sediment transport as the tide moves in and out, and it forages only during the daytime low tide. Predation by birds appears to be a major constraint as well (Zimmer-Faust 1989). Thus it seems that the costs of sensory mechanisms and time in selecting patches may outweigh the gains. The solution used to maintain rate of gain is geometrically very similar to that employed by echiurans (Fig. 7.4b). The crabs scuttle in swaths extending from the central burrow to a set distance, take a step forward and forage back, repeating the process around the compass. These swaths proceed either clockwise or anticlockwise, but are timed and spaced so as to cover the full circle in the one daytime low tide. The distance moved from the burrow thus is the minimum possible for the area of sand foraged. This foraging pattern represents an interesting variant on Richter's (1928) algorithm, and one that Zimmer-Faust (1987) documents to be approximately optimal. It would be interesting to compare swath geometries (i.e. their aspect ratios and numbers around the circle) of the crabs with those of echiurans.

Overall, there are no clear violations of foraging theory in documented examples of patch exploitation by deposit feeders. In few cases, however, have rules of thumb or 'satisficing' (doing well enough but not optimally cf. Stephens & Krebs 1986) been tested explicitly against truly optimal solutions. Such tests would be instructive if for no other reason than to see how close the simple algorithms obvious to Richter (1928, cited in Raup & Seilacher 1969) come to optimality.

DIET SELECTION

Diet selection also reflects the need for means to collect material quickly in large volumes. Taghon's (1988) results discussed above show that there is no great penalty in ingesting food of low food value. Thus diet selection should be on the basis of opportunity lost (*sensu* Stephens & Krebs 1986) by not selecting high-quality items rather than on the basis of cost of processing poor items. If selecting high-quality items takes too much time, such selection results in loss of the opportunity to ingest a greater mass of food of lesser quality with potentially greater rate of absorptive gain. These volume and time constraints act on diet choice to produce the nearly universal observation of partial preference in particle selection. Most species show a bias toward smaller particles, at least down to the range of 0–10 μm (e.g. Whitlatch 1980) yet ingest nearly the full spectrum of particle sizes below the morphological limit of mouth size. A few species show strong selection for larger particles (e.g. Whitlatch 1974), while others show little apparent selection. There are many reports of non-selective ingestion, but one must be careful to distinguish failure to find selection from convincing demonstration that selection does not occur. Since non-selectivity is the null hypothesis in virtually all these examinations, failure to reject it is not convincing evidence of the absence of selectivity without a formal test of statistical power. Thus most findings of apparent non-selectivity cannot be distinguished statistically from weakly selective feeding. The rates at which deposit feeders are constrained to feed may, however, make strong selection difficult or impossible in settings where sediment transport does not frequently renew the supply of particles.

The impact of feeding rate on diet selection is perhaps most evident when feeding rates are expressed as number of particles ingested per second. The small marine deposit feeder *Corophium* (an amphipod), for example, feeds at about 30 particles sec[-1] (Miller 1984). Self and Jumars (unpublished) estimate that a 30-cm long *Parastichopus californicus* (a holothuroid) when feeding on even relatively coarse sand (125 μm median grain diameter) ingests 10^3 particles sec[-1]. The point of this quantification is to suggest that choice of individual particles may be difficult if not impossible. At 30–1000 particles ingested sec[-1] it seems unlikely that evaluation routinely can be made particle by particle. Rather, mechanical means are becoming apparent that allow moderate bias toward smaller, less dense (g cm[-3]) or otherwise richer particles without greatly slowing the rate of ingestion. Particle selection can occur upon collection, after collection but before ingestion, or even after ingestion via selective gut passage.

The most widespread method of particle collection in deposit feeding is via mucous adhesives, often coupled with ciliary transport of the layer of mucus in which the particles are ensnared or muscular withdrawal or inversion of the sticky surface for ingestion. From 1 to 10^2 tentacles may be employed to enlarge the collection area with less predation risk than extending the whole body, but adhesion can act via the pharynx alone, as Jumars *et al.* (1982) documented in tentacle feeders that had lost their tentacles. Adhesion is used in particle collection by polychaetes, hydrobiid snails, protobranch bivalves, echiurans, sipunculans, holothuroids, echinoids, asteroids, enteropneusts, i.e. by nearly all soft-bodied deposit feeders and also by hard-bodied deposit feeders having soft appendages.

Mechanisms of contact of tentacle and body surfaces with particles have an inherent bias toward larger particles in the deposit (Jumars *et al.* 1982; Whitlatch 1989; Telford 1990). The bias is in direct proportion to particle radius if the collection device is modelled as planar but can approach the square of the radius in the extreme of a zero-width 'line' applied to sediments. Hentschel (in preparation) has predicted and shown that small, tentacle-feeding individuals, whose tentacle widths approach the diameters of particles that they encounter, experience an even greater bias toward large particles than do larger individuals of the same species. Jumars *et al.* (1982) demonstrated with analogue experiments that for spherical particles (and arguably for particles of any shape so long as shape remains constant across size classes cf. Weibel 1963) and for $d_i < d_j$ where d_i is diameter of the ith particle size class, n_i is the number of encounters of the ith particle size class and Q_i is the number of beads of the ith size class per unit of sediment volume:

$$\frac{n_i}{n_j} \leq \frac{d_i Q_i}{d_j Q_j} \tag{7.4}$$

Coming from the stereology of thin sections, this formulation under-estimates the large-particle bias of appressing a flat plane on the sediment surface, from which larger particles will protrude further. This protrusion bias can be overcome by some combination of a layer of mucus (especially if it is thick enough to span the difference among particle heights), ciliation, papillation or rugosity of the collecting surface.

Rather than showing consistent bias towards large particles, however, mucous adhesive-using deposit feeders were found to ingest smaller spheres preferentially down to about 10 μm, below which selectivity again fell (Jumars *et al.* 1982; Self & Jumars 1988). It is important to stress that the feed particles were clean glass and plastic beads of **no** food value. This unrealistic choice of particles was provided to allow more precise analysis of the mechanical workings of the selection process by eliminating complex geometries and variations in shape and limiting differences in surface properties. For this purpose spheres work all the better because they are out of the realm of organism experience. Little post-contact rejection behaviour was seen in the adhesive-feeding animals in these experiments. Jumars *et al.* (1982) proposed that adhesive failure to retain particles that were too heavy per unit of surface area contacted could account for the observed pattern of selection. Thus probability of contact multiplied by the conditional probability of retention given contact would give the observed, unimodal selection curves peaking at about 10 μm. As predicted from their monotonically decreasing weight per unit of surface area, particles of decreasing specific gravity were ingested preferentially. Selection for large particles thus easily can be explained as a consequence of strong adhesives.

While it has not yet been tested rigorously, this two-step model still appears consistent with published results. More recently, Guieb, Jumars and Self (in preparation) have extended similar experiments with the polychaete *Pseudopolydora kempi japonica* to size-graded, natural sediment grains. As expected, natural grains with rougher surface textures than glass beads are easier to pick up (more surface area of adhesive per contact), shifting the most preferred grain sizes upward by a factor of 10 over the most preferred sizes of spheres. In addition, one size class at a time was coated selectively with a bacterial monoculture. Through a set of internal controls (glass beads of varying sizes interspersed with the natural grains), it was concluded that the animals were able mechanically to select preferentially the coated grains but showed no behavioural flexibility to change the strength of their adhesives to enhance retention of particular size classes rich in food.

This adhesive collection machine is elegantly simple and efficient; heavy, food-poor particles incur no cost of transport since they are not picked up, and great rates of collection are achievable since individual particles need not be evaluated. It is sensitive, however, to the ambient grain-size distribution (cf. Equation 7.3) in terms of selectivity achievable and would appear to be sensitive to cohesion among sediment grains. Sticking together of grains must reduce selective ability for individual components of the aggregate, but the effect need not be entirely negative. *Pseudopolydora kempi japonica*, for example, ceases to feed when there are no unattached grains at the sediment surface, while the sympatric tentacle feeder *Hobsonia florida* continues to feed by-pulling grains from the bed so long as overlying water remains (Jumars & Self 1986). Although *H. florida* thus achieves a greater average rate of intake, *P. kempi japonica* presumably obtains richer, fresher organic matter. Similarly, adhesive feeders show great diversity in the degree to which particles are actively rejected after adhesive capture. By comparison with other spionid polychaetes (e.g. Dauer *et al.* 1981), *P. kempi japonica* shows comparatively little rejection after capture, again consistent with its pickup of unattached grains.

The next most prevalent means of particle pickup to adhesion is simple scooping, as seen in many crustaceans. Scooping up a volume of sediments eliminates or at least reduces the bias towards collection of larger particles. It is nearly always accompanied, however, by some mechanism for post-capture, pre-ingestive rejection. Post-capture sorting is perhaps most evident in the intertidal ocypodid crabs *Uca* and *Scopimera* that form discrete boluses of rejected material that volumetrically and gravimetrically grossly exceed the amount of material that is accepted into the alimentary tract for digestion. Their sorting is accomplished by elutriation and flotation. The mechanism has long been known qualitatively (Altevogt 1957; Miller 1961; Fielder 1970), but only recently has received quantitative mechanical analysis (Robertson & Newell 1982). It is clear, however, that great concentration of organic matter for ingestion is achieved, particularly by those species inhabiting sands (Robertson & Newell 1982; Zimmer-Faust 1987). The requirement of a chamber for elutriation limits this mechanism to relatively large-bodied species. In the much smaller amphipod *Corophium*, several sets of experiments document preference for smaller particles (e.g. Fenchel *et al.* 1975; Self & Jumars 1988). Miller (1984) suggested that at least part of the rejection is mechanical. He presented a Poisson model in which relatively coarse setae are used to 'rake' coarse particles into a rejection stream. The probability, P_0, that a particle would not be dislodged (rejected) by n seta-particle contacts, where $p = b \times d$ is the probability of dislodgement for each seta-particle contact, b is a coefficient of proportionality and d is particle diameter, is given by:

$$P_0 = e^{-bdn} \tag{7.5}$$

This model is consistent with Miller's (1984) observed linear dependence of selectivity on diameter. When coupled with a reduced ability to retain the smallest particles (Miller 1984) it is capable of producing a unimodal selection curve (Self & Jumars 1988) very similar to those of adhesive feeders. *Corophium* is also unusually efficient at choosing protein-coated glass beads from uncoated ones (Taghon 1982), suggesting some involvement of adhesion or more specific and behavioural, chemical selection. These crustacean mechanisms would appear to be less sensitive to ambient size–frequency distributions and to cohesion than are adhesive mechanisms, perhaps contributing to the increasing prevalence of crustacean deposit feeders in the abyss (Jumars & Gallagher 1982). Selection by crustacean mechanisms for a particular particle type or size also appears to be less sensitive than adhesive mechanisms to the composition of ambient particle mixtures (cf. Equation 7.3 vs. 7.4 and the results of Robertson & Newell (1982) and Miller (1984).

The only other major class of collection methods is suction, seen in the tellinid bivalves. It also appears to be relatively non-selective, i.e. to have sufficient power to draw in any particle that does not exceed the inhalant siphon's inside diameter (Self & Jumars 1988). As with scooping, pre-ingestive sorting after collection is the rule and is effective for both specific gravity and particle size (Hughes 1975; Hylleberg & Gallucci 1975). Again, the ciliary tracts involved have been described qualitatively, but neither a quantitative mechanical analysis of the sorting mechanism nor a suggested dependence on particle size like Equation 7.3 or 7.4 has yet been published. The net result, however, is ingestion of particle spectra remarkably similar to those achieved with the other mechanisms (Self & Jumars 1988). Suction mechanisms of deposit feeding are not known from deep water.

It is easy to overlook some 'passive' means of selection. Since sediment grain sizes reflect extreme transport events more than they do the mean condition, there often is a thin veneer of recently deposited material or material being barely transported by the mean condition. A great advantage of surface deposit feeding is spatial access to this material. This material of low specific gravity and small grain size is available without any need for sorting devices. Use of external physical processes to accomplish selection is less well known in subsurface deposit feeders, but the arrangement of sediments in trace fossils of subsurface deposit feeder activity provides abundant evidence of grain-size selective slumping (Seilacher 1986), and there is observational evidence on at least one species (Kudenov 1978) that such slumping is used to obtain finer grain sizes for ingestion.

Hydrobiid snails (Lopez & Kofoed 1980) and corophiid amphipods (Nielsen and Kofoed 1982) present another interesting variant on selection by diverging from deposit feeding *sensu strictu*. Over much of the sedimentary particle size range they show selection spectra and partial preferences much like those of other deposit feeders. For grains above about 40 μm in diameter, however, they rasp off surface coatings for ingestion rather than ingesting grains. At this extreme, handling particles one at a time must be worthwhile. The gut space saved for ingestion of smaller particles, including the scrapings, must repay the time and energy costs. That this behaviour is prominent in small deposit feeders, for which an ingested large grain represents a substantial fraction of gut volume, again underscores the volume and throughput constraints of deposit feeding. It would be interesting to compare the limits on grain sizes ingested with corresponding limits on seed size in frugivores. In several ways, eating of fruit with large seeds represents an interesting analogy with deposit feeding.

Although chemical cues stimulating enhanced feeding rate are known, chemicals that induce active behaviours (e.g. ciliary reversals) that lead to preferential retention or rejection of particles coated with them have not been identified. Taghon (1982) showed enhanced selection of protein-coated glass beads over clean glass beads in most of the species that he studied. Without documentation of behavioural change, however, these data could be interpreted simply as a mechanical consequence of the greater 'stickiness' of coated beads. Observations do show behavioural rejection of individual particles and of particle boluses, the ciliary field of an entire tentacle sometimes being reversed, but the chemical or physical cues have not been identified.

There are trends towards less selectivity in larger-bodied species and subsurface deposit feeders (Self & Jumars 1988), for which rates of supply relative to feeding rates may limit selectivity. More generally, effects on rate of absorption as modelled by Dade *et al.* (1990), give a means for estimating maximal costs of selection: selection should occur only to the extent that the time taken to achieve it does not cause greater loss, in terms of the rate of absorption, than would be gained by that selection. Individuals feeding at maximal rates, i.e. on food of intermediate quality, thus may forego particle selection to a greater extent than individuals feeding on either very low- or very high-quality food.

One means of diet selection not often considered is sorting in the gut. The time constraints imposed by plug flow (Chapter 3) can be relieved to some extent by selective retention of materials that continue to provide absorptive gain and selective passage of food-poor and rapidly utilized materials. Such selective retention is one clear advantage of intracellular digestion, seen notably in some molluscs. There is evidence in some deposit-feeding species for sorting in the gut lumen on the basis of physical properties (Self & Jumars 1978; Penry 1989) and correlated biochemical properties (Kofoed *et al.* 1989). Recent results of Decho and Luoma (1991) suggest that intracellular digestion in some deposit feeders may provide access to food with relatively slow digestive kinetics, though volumetric throughput rates cannot be large. The slowed kinetics of throughput via intracellular digestion may also help to explain the molluscan outliers in Fig. 7.2 (page 131). A particularly interesting question then becomes the means by which particles are selected for this digestive treatment.

Conversely, there are other species, such as *Capitella* sp. I (Forbes & Lopez 1990) that bind their particulate food almost immediately upon ingestion into pellets that preclude particle sorting in the gut. Yet to be determined in either kind of species is the relative passage rate of fluid and particulate phases, though one qualitative experiment (Jumars unpublished) suggests, as expected from experience with larger animals in which tracers are more easily employed, a longer residence time of the fluid. *Pseudopolydora kempi japonica* individuals were each fed (sequentially) anaerobic sediments, Sephadex® beads saturated with tetrazolium salts, and anaerobic sediments again. The salts are soluble until they are reduced, when they precipitate irreversibly. Particles travel in plug flow in *P. kempi japonica* (Jumars & Self 1986), providing a clear demarcation of sediments and chromatography beads. A front of precipitation often was evident anterior to the plug of Sephadex® but was not found posterior of it, implying generally anteriorward transport of fluid relative to particles. These results are consistent both with the greater fluid than particulate residence times known for large mammals (e.g. Van Soest 1982) and with the ongoing work of Mayer *et al.* (in preparation), who find that solubilization of large polypeptides from marine sediments occurs faster than their hydrolysis to assimilable oligomer size, making longer fluid retention advantageous.

CONCLUSIONS

These various observations support the notion of rate (i.e. time) constraints as strong determinants of diet selection in deposit feeders and suggest that deposit feeding may represent the natural extreme of chronic limitation by dilute diet and consequent need for rapid, nearly continuous feeding. Time and sensory costs can preclude selection of patches and usually do yield partial preference for poor foods. There is an element of *déjà vu* suggesting caution, however, in the conclusion that individual particles probably are not evaluated for ingestion. This notion was prevalent in early studies of suspension feeders but has been tempered by direct observation (e.g. Price *et al.* 1982 and Chapter 6). Such direct observation is clearly needed in a diversity of deposit feeders, but most notably the subsurface ones, where such observation is the most difficult.

For very few species of deposit feeders is there an organized body of information on the entire foraging strategy. Rather, there are isolated studies of particle or patch selection or of feeding rate. The two most conspicuous exceptions are species from two genera, *Uca* and *Hydrobia*, both of which are intertidal and specialize on benthic diatoms as food. The various studies by Robertson and coworkers cited above show that a sand-dwelling species of *Uca* selects primarily on the basis of patches – and *Hydrobia* appears to do the same (Forbes & Lopez 1986). This generalization fails to hold, however, even for other mobile, intertidal deposit feeders. Knowledge of foraging strategies of subtidal deposit feeders is fragmentary. For few living subsurface deposit feeders – intertidal or subtidal – are frequencies and patterns of movement known. The situation is arguably better for the fossil record.

This dearth of integrated information is matched, again with one exception from Robertson's studies (as extended by Weissburg & Zimmer-Faust 1991), by particular lack of information on chemical cues that initiate ingestion and modulate its rates in deposit feeders. Given the demonstrable profitability of such information in understanding the foraging of insects (Bernays 1985), this lack is all the more glaring. Subsurface deposit feeders are well supplied with chemosensory equipment (e.g. the nuchal organs of polychaetes) whose capabilities might give good clues to the kinds of patches that subsurface deposit feeders prefer and thus to the resources that support them.

In summary, marine deposit feeding is seen to be an extreme digestive strategy, entailing rapid throughput of organically dilute, largely indigestible material. Marine deposit feeders contrast markedly with fermenters (Chapter 3) that extract energy from refractory but ultimately digestible material and that hence have much longer gut residence times. The constraint to operate at high volumetric throughput restricts deposit feeders to relatively rapidly digestible material, limits minimal sizes of deposit feeders, selects for simple but effective means of patch utilization and results in the dietary inclusion of particles of low or zero food value. Machine-like methods of particle collection and sorting used by deposit feeders from diverse taxa converge in showing peak selection of lower excess density (density of the particle minus density of the fluid). The sorting 'machinery' thus converges evolutionarily on physical characteristics associated with geophysical transport and renewal of deposits, i.e. on newly arriving material.

REFERENCES

Altevogt R. (1957) Untersuchungen zur Biologie, Ökologie und Physiologie Indischer Winkerkrabben. *Z. Morphol. Ökol. Tiere* **46**, 1–110.

Bernays E.A. (1985) Regulation of feeding behaviour. In *Comprehensive Insect Physiology, Biochemistry and Pharmacology. Vol. 4. Regulation: Digestion, Nutrition, Excretion.* (ed. by G.A. Kerkut & L.I. Gilbert), pp. 1–32. Pergamon Press, Oxford.

Bignell D.E. (1984) The arthropod gut as an environment for microorganisms. In *Invertebrate–Microbial Interactions* (ed. by J.M. Anderson, A.D.M. Rayner and D.W.H. Walton), pp. 205–27. Cambridge University Press, Cambridge.

Billett D.S.M., Lampitt R.S., Rice A.L. & Mantoura R.F.S. (1983) Seasonal sedimentation of phytoplankton to the deep-sea benthos. *Nature* **302**, 520–2.

Butman C.A. (1987) Larval settlement of soft-sediment invertebrates: the spatial scales of pattern explained by active habitat selection and the emerging rôle of hydrodynamical processes. *Oceanogr. Mar. Biol., Ann. Rev.* **25**, 113–65.

Butman C.A., Grassle J.P. & Webb C.M. (1988) Substrate choices made by marine larvae in still water and in a flume flow. *Nature* **333**, 771–3.

Cammen L.M. (1980) Ingestion rate: an empirical model for aquatic deposit feeders and detritivores. *Oecologia* **44**, 303–10.

Cammen L.M. (1982) Effect of particle size on organic content and microbial abundance within four marine sediments. *Mar. Ecol. Progr. Ser.* **9**, 273–80.

Cammen L.M. (1987) Polychaeta. In *Animal Energetics. Vol. 1. Protozoa through Insecta* (ed. by T.J. Pandian & F.J. Vernberg), pp. 217–60. Academic Press, San Diego.

Cammen L.M. (1989) The relationship between ingestion rate of deposit feeders and sediment nutritional value. In *Ecology of Marine Deposit Feeders* (ed. by G.R. Lopez, G.L. Taghon & J.S. Levinton), pp. 201–22. Springer Verlag, Berlin.

Dade W.B., Jumars P.A. & Penry D.L. (1990) Supply-side optimization: Maximizing absorptive rates. In *Behavioural Mechanisms of Food Selection* (ed. by R.N. Hughes), *NATO ASI series, vol. G 20*, pp. 531–56. Springer Verlag, Berlin.

Dale N.G. (1974) Bacteria in intertidal sediments: Factors related to their distribution. *Limnol. Oceanogr.* **19**, 509–18.

Dauer D.M., Maybury C.A. & Ewing R.M. (1981) Feeding behavior of several spionid polychaetes from the Chesapeake Bay. *J. Exp. Mar. Biol. Ecol.* **54**, 21–38.

Decho A.W. & Luoma S.N. (1991) Time-courses in the retention of food material in the bivalves *Potamocorbula amurensis* and *Macoma balthica*: significance to the absorption of carbon and chromium. *Mar. Ecol. Progr. Ser.* **78**, 303–14.

DeFlaun M.F. & Mayer L.M. (1983) Relationships between bacteria and grain surfaces in intertidal sediments. *Limnol. Oceanogr.* **28**, 873–81.

Demment, M.W. & Longhurst W.M. (1987) Browsers and grazers: Constraints on feeding ecology imposed by gut morphology and body size. *Proc. IV International Conference on Goats* (ed. by D.P. Santana, A.G. da Silva & W.C. Foote), pp. 989–1004. Departamento de Difusao de Tecnologia, Brazilia, Brazil.

Eckman J.E. & Nowell A.R.M. (1984) Boundary skin friction and sediment transport about an animal-tube mimic. *Sedimentology* **31**, 851–62.

Emerson C. & Grant J. (1991) Control of soft-shell clam (*Mya arenaria*) recruitment on intertidal sandflats by bedload transport. *Limnol. Oceanogr.* **36**, 1288–300.

Fauchald K. & Jumars P.A. (1979) The diet of worms: A study of polychaete feeding guilds. *Oceanogr. and Mar. Biol., Ann. Rev.* **17**, 193–284.

Fenchel T., Kofoed L.H. & Lappalainen A. (1975) Particle size-selection of two deposit feeders: the amphipod *Corophium volutator* and the prosobranch *Hydrobia ulvae*. *Mar. Biol.* **30**, 119–28.

Fielder D.R. (1970) The feeding behaviour of the sand bubbler crab *Scopimeria inflata* (Decapoda, Ocypodidae). *J. Zool.* **160**, 35–49.

Fong, W. & Mann K.H. (1980) Role of gut flora in the transfer of amino acids through a marine food chain. *Can. J. Fish. Aquat. Sci.* **37**, 88–96.

Forbes V.E. & Lopez G.R. (1986) Changes in feeding and crawling rates of *Hydrobia truncata* (Prosobranchia: Hydrobiidae) in response to sedimentary chlorophyll-*a* and recently egested sediment. *Mar. Ecol. Progr. Ser.* **33**, 287–94.

Forbes T.L. & Lopez G.R. (1990) Ontogenetic changes in individual growth and egestion rates in the deposit-feeding polychaete *Capitella* sp. 1. *J. Exp. Mar. Biol. Ecol.* **143**, 209–20.

Fuller C.M., Butman C.A. & Conway N.M. (1988) Periodicity in fecal pellet production by the capitellid polychaete *Mediomastus ambiseta* throughout the day. *Ophelia* **29**, 83–91.

Gallagher E.D., Gardner G.B. & Jumars P.A. (1990) Competition among the pioneers in a seasonal soft-bottom benthic succession: Field experiments and analysis of the Gilpin–Ayala model. *Oecologia* **83**, 427–42.

Gibbs P.E., Bryan G.W. & Ryan K.P. (1981) Copper accumulation by the polychaete *Melinna palmata*: An antipredation mechanism? *J. Mar. Biol. Assoc. U.K.* **61**, 707–22.

Grant J. (1983) The relative magnitude of biological and physical sediment reworking in an intertidal community. *J. Mar. Res.* **41**, 673–89.

Hargrave B.T. & Phillips G.A. (1977) Oxygen uptake of microbial communities on solid surfaces. In *Aquatic Microbial Communities* (ed. by J. Cairns), pp. 545–687. Garland Press, N.Y.

Hoppe P.P. (1977) Rumen fermentation and body weight in African ruminants. In *Proc. XIIIth International Congress of Game Biologists* (ed. by T.J. Peterle). The Wildlife Society, Washington, D.C.

Hughes T.G. (1975) The sorting of food particles by *Abra* sp. (Bivalvia: Tellinacea). *J. Exp. Mar. Biol. Ecol.* **20**, 137–56.

Hylleberg J. & Gallucci V.F. (1975) Selectivity in feeding by the deposit-feeding bivalve *Macoma nasuta*. *Mar. Biol.* **32**, 167–78.

Illius A.W. & Gordon I.J. (1990) Constraints on diet selection and foraging behaviour in mammalian herbivores. In *Behavioural Mechanisms of Food Selection* (ed. by R.N. Hughes), *NATO ASI series, vol. G 20*, pp. 369–93. Springer Verlag, Berlin.

Jumars P.A. & Gallagher E.D. (1982) Deep-sea community structure: Three plays on the benthic proscenium. In *The Environment of the Deep Sea* (ed. by W.G. Ernst & J.G. Morin), pp. 217–55. Prentice-Hall, Englewood Cliffs, N.J.

Jumars P.A. & Self R.F.L. (1986) Gut-marker and gut-fulness methods for estimating field and laboratory effects of sediment transport on ingestion rates of deposit-feeders. *J. Exp. Mar. Biol. Ecol.* **98**, 293–310.

Jumars P.A., Self R.F.L. & Nowell A.R.M. (1982) Mechanics of particle selection by tentaculate deposit feeders. *J. Exp. Mar. Biol. Ecol.* **64**, 47–70.

Jumars P.A., Newell R.C., Angel M.V., Fowler S.W., Poulet S.A., Rowe G.T. & Smetacek V. (1984) Detritivory. In *Flows of Material and Energy in Marine Ecosystems* (ed. by M.J.R. Fasham), pp. 685–93. Plenum Press, N.Y.

Jumars P.A., Mayer L.M., Deming J.W., Baross J.A. & Wheatcroft R.A. (1990) Deep-sea deposit-feeding strategies suggested by environmental and feeding constraints. *Phil. Trans. Roy. Soc. London. A* **331**, 85–101.

Kofoed L., Forbes V. & Lopez G. (1989) Time-dependent absorption in deposit feeders. In *Ecology of Marine Deposit Feeders* (ed. by G.R. Lopez, G.L. Taghon & J.S. Levinton), pp. 129–48. Springer Verlag, Berlin.

Kudenov J.R. (1978) The feeding ecology of *Axiothella rubrocincta* (Johnson) (Polychaeta: Maldanidae). *J. Exp. Mar. Biol. Ecol.* **31**, 209–21.

Levinton J.S. (1971) Control of tellinacean (Mollusca: Bivalvia) feeding behavior by predation. *Limnol. Oceanogr.* **16**, 660–2.

Levinton J.S. & Lopez G.R. (1977) A model of renewable resources and limitation of deposit-feeding benthic populations. *Oecologia* **31**, 177–90.

Longbottom M.R. (1970) The distribution of *Arenicola marina* (L.) with particular reference to the effects of particle size and organic matter of the sediments. *J. Exp. Mar. Biol. Ecol.* **5**, 138–57.

Lopez G.R & Kofoed L.H. (1980) Epipsammic browsing and deposit-feeding in mud snails (Hydrobiidae). *J. Mar. Res.* **38**, 585–99.

Lopez G.R. & Levinton J.S. (1987) Ecology of deposit-feeding animals in marine sediments. *Quart. Rev. Biol.* **62**, 235–60.

Lopez G.R., Taghon G.L. & Levinton J.S. (eds) (1989) *Ecology of Marine Deposit Feeders*. Springer Verlag, Berlin.

Mangel M. & Clark C.W. (1988) *Dynamic Modeling in Behavioral Ecology*. Princeton University Press, Princeton, N.J.

Mayer L.M. (1989) The nature and determination of non-living sedimentary organic matter as a food source for deposit feeders. In *Ecology of Marine Deposit Feeders* (ed. by G.R. Lopez, G.L. Taghon & J.S. Levinton), pp. 98–113. Springer Verlag, Berlin.

Mayer L.M., Jumars P.A., Taghon G.L., Macko S.A. and Trumbore S. Low-density particles as nitrogenous foods for benthos. *J. Mar. Res.*, in review.

Meadows P.S.M. (1964) Experiments on substrate selection by *Corophium* species: Films and bacteria on sand particles. *J. Exp. Biol.* **41**, 499–511.

Miller D.C. (1961) The feeding mechanisms of fiddler crabs, with ecological considerations of feeding adaptations. *Zoologica* **46**, 89–101.

Miller D.C. (1984) Mechanical post-capture particle selection by suspension- and deposit-feeding *Corophium*. *J. Exp. Mar. Biol. Ecol.* **82**, 59–76.

Miller D.C. & Jumars P.A. (1986) Pellet accumulation, sediment supply, and crowding as determinants of surface deposit-feeding rate in *Pseudopolydora kempi japonica* Imajima and Hartman (Polychaeta: Spionidae). *J. Exp. Mar. Biol. Ecol.* **99**, 1–17.

Miller D.C. & Sternberg R.W. (1988) Field measurements of the fluid and sediment-dynamic environment of a benthic deposit feeder. *J. Mar. Res.* **46**, 771–96.

Nielsen M.V. & Kofoed L.H. (1982) Selective feeding and epipsammic browsing by the deposit-feeding amphipod *Corophium volutator*. *Mar. Ecol. Progr. Ser.* **10**, 81–8.

Newell R.C. (1965) The role of detritus in the nutrition of two marine deposit feeders, the prosobranch *Hydrobia ulvae* and the bivalve *Macoma balthica*. *Proc. Zool. Soc. Lond.* **144**, 25–45.

Nichols F.H., Cacchione D.A., Drake D.E. & Thompson J.K. (1989) Emergence of burrowing urchins from California continental shelf sediments. *Estuar. Coastal Shelf Sci.* **29**, 171–82.

Nowell A.R.M., Jumars P.A. & Fauchald K. (1984) The foraging strategy of a subtidal and deep-sea deposit feeder. *Limnol. Oceanogr.* **29**, 645–9.

Nowell A.R.M., Jumars P.A., Self R.F.L. & Southard J.B. (1989) The effects of sediment transport and deposition on infauna: Results obtained in a specially designed flume. In *Ecology of Marine Deposit Feeders* (ed. by C.R. Lopez, G.L. Taghon & J.S. Levinton), pp. 247–68. Springer Verlag, Berlin.

Ohta S. (1984) Star-shaped feeding traces produced by echiuran worms on the deep-sea floor of the Bay of Bengal. *Deep-Sea Res.* **31**, 1415–32.

Papentin F. (1973) A Darwinian evolutionary system. III. Experiments on the evolution of feeding patterns. *J. Theor. Biol.* **39**, 431–45.

Penry D.L. (1989) Tests of kinematic models for deposit-feeder guts: Patterns of sediment processing by *Parastichopus californicus* (Stimpson) (Holothuroidea) and *Amphicteis scaphobranchiata* Moore (Polychaeta). *J. Exp. Mar. Biol. Ecol.* **128**, 127–46.

Penry D.L. & Jumars P.A. (1987) Modeling animal guts as chemical reactors. *Am. Nat.* **129**, 69–96.

Penry D.L. & Jumars P.A. (1990) Gut architecture, digestive constraints and feeding ecology of deposit-feeding and carnivorous polychaetes. *Oecologia* **82**, 1–11.

Phillips N.W. (1984) Compensatory intake can be consistent with an optimal foraging model. *Am. Nat.* **123**, 867–72.

Plante C.J., Jumars P.A. & Baross J.A. (1990) Digestive associations between marine detritivores and bacteria. *Ann. Rev. Ecol. Syst.* **21**, 93–127.

Price H.J., Paffenhöfer G.-A. & Strickler J.R. (1982) Modes of cell capture in calanoid copepods. *Limnol. Oceanogr.* **28**, 116–23.

Raup D.M. & Seilacher A. (1969) Fossil foraging behavior: Computer simulation. *Science* **166**, 994–5.

Reimers C.E. (1989) Control of benthic fluxes by particulate supply. In *Productivity of the Ocean: Present and Past* (ed. by W.H. Berger, V.S. Smetacek & G. Wefer), pp. 217–33. John Wiley & Sons, Chichester.

Robertson J.R. & Newell S.Y. (1982) Experimental studies of particle ingestion by the sand fiddler crab *Uca pugilator* (Bosc). *J. Exp. Mar. Biol. Ecol.* **59**, 1–21.

Robertson J.R., Bancroft K., Vermeer G. & Plaisir K. (1980) Experimental studies on the foraging behavior of the sand fiddler crab *Uca pugilator* (Bosc, 1802). *J. Exp. Mar. Biol. Ecol.* **44**, 67–83.

Robertson J.R., Fudge J.A. & Vermeer G.K. (1981) Chemical and live feeding stimulants of the sand fiddler crab, *Uca pugilator* (Bosc). *J. Exp. Mar. Biol. Ecol.* **53**, 47–64.

Scheibling R.E. (1981) Optimal foraging movements of *Oreaster reticulatus* (L.) (Echinodermata: Asteroidea). *J. Exp. Mar. Biol. Ecol.* **51**, 173–85.

Seilacher A. (1986) Evolution of behavior as expressed in marine trace fossils. In *Evolution of Animal Behavior* (ed. by M.H. Nitecki & J.A. Kitchell), pp. 62–87. Oxford University Press, New York.

Self R.F.L. & Jumars P.A. (1978) New resource axes for deposit feeders? *J. Mar. Res.* **36**, 627–41.

Self R.F.L. & Jumars P.A. (1988) Cross-phyletic patterns of particle selection by deposit feeders. *J. Mar. Res.* **46**, 119–43.

Stephens D.W. & Krebs J.R. (1986) *Foraging Theory.* Princeton University Press, Princeton N.J.

Taghon G.L. (1981) Beyond selection: Optimal ingestion rate as a function of food value. *Am. Nat.* **118**, 202–14.

Taghon G.L. (1982) Optimal foraging by deposit-feeding invertebrates: Roles of particle size and organic coating. *Oecologia* **52**, 295–304.

Taghon G.L. (1988) The benefits and costs of deposit feeding in the polychaete *Abarenicola pacific. Limnol. Oceanogr.* **33**, 1166–75.

Taghon G.L. & Jumars P.A. (1984) Variable ingestion rate and its role in optimal foraging behavior of marine deposit feeders. *Ecology* **65**, 549–58.

Taghon G.L. & Greene R.R. (1990) Effects of sediment-protein concentration on feeding and growth rates of *Abarenicola pacifica* Healy et Wells (Polychaeta: Arenicolidae). *J. Exp. Mar. Biol. Ecol.* **136**, 197–216.

Taghon G.L., Self R.F.L. & Jumars P.A. (1978) Predicting particle selection by deposit feeders: A model and its implications. *Limnol. Oceanogr.* **23**, 752–9.

Telford M. (1990) Computer simulation of deposit-feeding by sand dollars and sea biscuits (Echinoidea: Clypeasteroidea) *J. Exp. Mar. Biol. Ecol.* **142**, 75–90.

Tenore K.R. & Hanson R.B. (1980) Availability of detritus of different types and ages to a polychaete macroconsumer, *Capitella capitata. Limnol. Oceanogr.* **25**, 553–8.

Valiela I., Koumjian L., Swain T., Teal J. & Hobbie J.E. (1979) Cinnamic acid inhibition of detritus feeding. *Nature* **280**, 55–7.

Van Soest P.J. (1982) *Nutritional Ecology of the Ruminants.* O & B Books, Corvallis, OR.

Watling L. (1988) Small-scale features of marine sediments and their importance to the study of deposit feeding. *Mar. Ecol. Progr. Ser.* **47**, 135–44.

Weibel E.R. (1963) Principles and methods for the morphometric study of the lung and other organs. *Lab. Invest.* **12**, 131–55.

Weissburg M. (1992) Sex and the single forager: gender-specific energy maximization strategies in the fiddler crab *Uca pugnax. Ecology,* in press.

Weissburg M.J. & Zimmer-Faust R.K. (1991) Ontogeny *versus* phylogeny in determining patterns of chemoreception: Initial studies with fiddler crabs. *Biol. Bull.* **181**, 205–15.

Weston D.P. (1990) Hydrocarbon bioaccumulation from contaminated sediment by the deposit-feeding polychaete *Abarenicola pacifica. Mar. Biol.* **107**, 159–69.

Whitlatch R.B. (1974) Food-resource partitioning the deposit feeding polychaete *Pectinaria gouldii. Biol. Bull.* **147**, 225–35.

Whitlatch R.B. (1980) Patterns of resource utilization and coexistence in marine intertidal deposit feeders. *J. Mar. Res.* **38**, 743–65.

Whitlatch R.B. (1989) On some mechanistic approaches to the study of deposit feeding in polychaetes. In *Ecology of Marine Deposit Feeders* (ed. by G.R. Lopez, G.L. Taghon & J.S. Levinton), pp. 291–308. Springer Verlag, Berlin.

Yager P.L., Nowell A.R.M. & Jumars P.A. (1993) Enhanced deposition to pits: a local food source for benthos. *J. Mar. Res.,* in press.

Zimmer-Faust R.K. (1987) Substrate selection and use by a deposit-feeding crab. *Ecology* **68**, 955–70.

Zimmer-Faust R.K. (1989) Foraging strategy of a deposit-feeding crab. In *Behavioural Mechanisms of Food Selection* (ed. by R.N. Hughes), *NATO ASI series, vol. G 20,* pp. 557–68. Springer Verlag, Berlin.

8: Diet Selection in Mammalian Herbivores: Constraints and Tactics

ANDREW W. ILLIUS and IAIN J. GORDON

INTRODUCTION

The purpose of this chapter is to review diet selection in ungulates, and in particular to consider the constraints that limit an animal's ability to recognize, select or utilize a particular food from amongst those apparently available to it. Intrinsic and extrinsic constraints are, respectively, properties of the animal and its environment which curtail the animal's foraging ability. For example, the animal's visual acuity, its minimum energy requirement and the abundance of nutritious but tiny leaves within a morass of thorn-scrub could each act as constraints on the diet selected. These constraints might perhaps require large animals with high energy requirements to turn away from such a food source once it had become depleted, while smaller animals might still find enough to survive on. This example introduces our main themes: the ways in which diet selection is constrained in herbivores; the influence of body size on these constraints; the decisions which animals make in selecting and exploiting food patches; and the consequences these have for the community ecology of ungulates.

IDENTIFYING THE CONSTRAINTS ON HERBIVORES

Niche selection and habitat utilization are the broadest expressions of the way animals resolve, both in the present and over evolutionary time, the conflicts between the need for food and the intrinsic and extrinsic constraints. Extrinsic constraints derive primarily from the nature of the food source, and interact with intrinsic constraints which are frequently related to an animal's body size, but also include psychological or behavioural constraints such as perception and discrimination. Modelling has shown that the mechanisms of niche selection are a function of two principal vegetation properties, abundance and quality, which interact with animal size through the processes governing food intake and digestion (Illius &

Gordon 1987, 1991). Controlled experimental tests of this are scarce, but the studies of Jarman and Sinclair (1979), Hobbs, Baker and Gill (1983) and Grant *et al.* (1985) give empirical support. In this chapter, it is argued that ungulate body sizes and feeding niches have evolved to occupy subsets of the vegetation quantity–quality matrix, and that animals have evolved the perceptual abilities to recognize the boundaries of these subsets. Within each subset, vegetation is heterogeneous in the properties which determine the rate of nutrient uptake and extraction by herbivores, and in the second half of the chapter attention is focussed on the decision-making processes involved in diet selection within the niche.

Before attempting to quantify the physical constraints, one must consider briefly what it is about herbivory that is supposed to distinguish it from other forms of existence such as carnivory. Herbivores inhabit a world in which at first sight the food plants are more or less continuously distributed in space and time. By comparison with a carnivorous predator, whose food is patchily distributed, highly nutritious and mobile, the herbivore confronts vegetation in which the nutrients are more evenly distributed, at low density, and whose accessibility is further reduced by structural and chemical properties of the vegetation. Variation in these properties, such as biomass density, extent of lignification and chemical content, is important in determining the potential value of a patch of vegetation, but may not present clear visual cues to animals. The morphology of the mouthparts and incisor dentition determine the extent to which an animal can prehend and ingest plant food items from within the spatial array of vegetation. Herbivores normally eat only parts of food plants, and the extent of partial consumption is the result of mechanical interactions, variation in feeding style and behavioural responses to the perceived physical properties of the vegetation. The dense admixture of plant parts (leaves, stems, twigs) and even of different plant species means that when the herbivore takes a bite of something, it is probably not quite sure what it is going to end up with (see Vivås, Saether & Anderson 1991). Furthermore, since the herbivore may be taking 10 000–40 000 bites from the vegetation in a day, the immediate significance of individual foraging choices for nutrient yield is likely to be lost. Vivås *et al.* (1991) argue that, since the herbivore must take so many bites in a day, strong selection will exist on identifying cues of optimal bite size. But stochastic variation in vegetation structure will certainly limit the animal's ability to optimize bite size and composition.

The herbivore's perception of the nutritional value of a food is unlikely to be instantaneous, since it cannot recognize most nutrients on

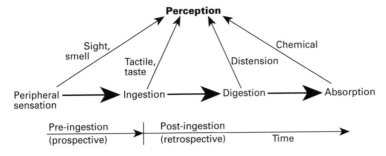

Fig. 8.1 Flow diagram of the processing of food and information.

consumption (Arnold 1981) and nutrient absorption may be delayed for several hours. This does not imply that herbivores are unselective or incapable of learning from the metabolic consequences of selection, especially if they are aversive (*see* Chapter 5). But it does suggest that they are unlikely to be able to perceive the more subtle differences in nutrient content that occur at a fine scale of resolution, such as during a succession of bites from broadly similar vegetation.

The herbivore's problem of how to gain and process information about its foraging choices is represented schematically in Fig. 8.1, showing the channels through which information is processed and the temporal separation of prospective faculties such as sight and smell, which precede consumption, and retrospective faculties such as touch, taste, gut distension and nutrient absorption.

Foraging theorists have generally ignored the influence of digestive constraints and nutrient absorption on foraging decisions by using animal models for which it can reasonably and conveniently be assumed that digestive efficiency is high and constant for alternative food items and across individuals. The size and spatio-temporal distribution of food items have been viewed as the dominant variables influencing foraging behaviour, with some exceptions (theoretical: Sibly 1981; Penry & Jumars 1987; herbivores: Jarman 1974; Owen-Smith & Novellie 1982; Belovsky 1978, 1986). The structural complexity of vegetation, with its highly variable composition and potential digestibility, necessitates the treatment of food quality as a variable equal to conventional variables in its capacity to influence choice. Semantically there may not seem to be much difference between patchily distributed nutriment (say, mealworms) and nutrients patchily distributed within a clump of vegetation. The difference is that in plants the nutrients are not only separated by space but also by indigestible chemical components, chiefly lignocellulose, and this

constraint pervades all herbivory. The claim that secondary plant compounds play the major role in mediating herbivory is enthusiastically pursued in the literature, but these compounds can surely be treated as just
one of a range of plant characteristics, albeit with occasionally spectacular
toxic effects.

QUANTIFYING CONSTRAINTS RELATED TO BODY SIZE

The way physical intrinsic constraints determine the category of plant
material the animal can use as food on the grounds of its quality
(digestible nutrient content) and abundance is now discussed. The body
size of mammalian herbivores has implications for the minimum quality
of food necessary for survival, and hence for the feeding niche selected.
This hypothesis was first proposed by Bell (1970) and Jarman (1974),
from a consideration of the allometric scaling of metabolic rate in ruminants, and is now termed the Bell–Jarman principle (Geist 1974). Bell
and Jarman argued that small antelope require high quality (low fibre)
food to satisfy their relatively high metabolic rates, assuming all antelope
are able to digest similar quantities of food relative to body mass, and
that this explained the differences in the diets consumed by African
ungulate guilds. An influential paper by Demment and Van Soest (1985)
explored these relationships, and expressed support for the principle by
inference from a number of lines of evidence. In particular, they argued
that large animals had a greater capacity to process and survive on poor
quality forages with slow fermentation rates, since energy requirements
scale with weight, W, as $W^{0.75}$ and gut contents scale isometrically with
W.

To achieve a better quantitative understanding of how food and
animal variables influence the rate of energy yield from the gut, a model
of digestion kinetics was developed to predict intake and digestion of
any specified forage by ruminants and hindgut-fermenting species (Illius
& Gordon 1991, 1992; *see also* Chapter 3). Intake in each meal was
defined as the amount of food required to refill the digestive tract after
the partial digestion and passage of previous meals. The clearance of
food from the digestive tract is the result of the differential rates of
digestion and passage of food components such as cell wall and cell contents. Passage rate is the inverse of retention time, which scales with
$W^{0.27}$ (Illius & Gordon 1991), in common with the scaling of the duration of other time-related physiological variables (Taylor 1980; Peters
1983). Hindgut fermenters have rates of passage which are nearly twice

Table 8.1 Details of the proportionate composition of foods chosen in the model of digestion kinetics to satisfy the energy requirements of a guild of ruminants

Food	Cell contents	Digestible cell wall	Indigestible cell wall	Digestion rate (h^{-1})	Potential digestibility
Grass 1	0.31	0.39	0.30	0.037	0.70
Grass 2	0.33	0.40	0.27	0.044	0.73
Browse	0.56	0.15	0.29	0.030	0.71
Forb	0.52	0.19	0.29	0.063	0.71

as fast as ruminants (Illius & Gordon 1992), because ruminants selectively retain large fibre particles in the rumen until digestion and physical comminution has reduced particle size to about 1 mm (Poppi, Minson & Ternouth 1981). This acts to delay passage and allow more extensive microbial digestion of a food fraction which the animal is otherwise incapable of digesting. The resulting energy yield from digestion can then be expressed as the multiple of the maintenance energy requirements the modelled animal can obtain from an abundant food eaten at the maximum rate allowed by clearance of digesta and residues from the gut.

The model was used to determine the different foods required by an imaginary size-structured guild of four African ruminants to satisfy exactly their maintenance requirements for energy (Table 8.1). The animals were: Thomson's gazelle, a mixed feeder of 20 kg body mass making highly selective use of forbs; impala, a mixed feeder using browse and weighing 60 kg; wildebeest, a grazer of 180 kg; and African buffalo, a grazer of 540 kg. Browse, in the form of leaves of shrubs, is characterized by a high proportion of cell solubles but poorly digestible cell walls; grass cell solubles' proportion declines to low levels with advancing maturity, but cell walls are more digestible at an equivalent phenological stage to browse; and forbs are intermediate in these respects, resembling browse more than grass (see Short, Blair & Segelquist 1974). The model was also used to compare hindgut fermenters with ruminants to quantify differences in digestive constraints. Figure 8.2a,b shows the effect of body size on the energy yield from the gut, relative to maintenance needs, when these foods are eaten at maximum daily intake. The results can be summarized as follows:

1 Hindgut fermenters can obtain more energy from an abundant food than ruminants because their higher rates of passage allow them to eat

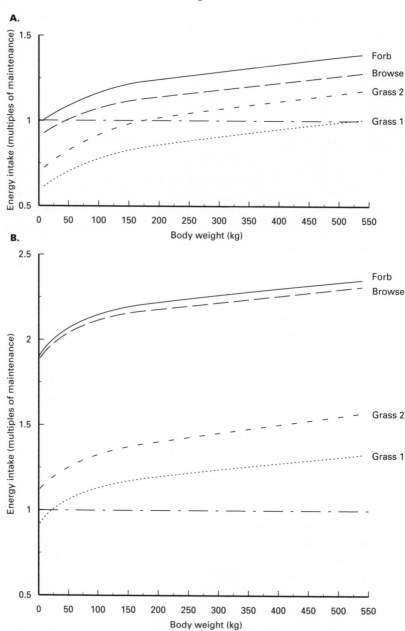

Fig. 8.2 Modelled relationship between body size and the ability of (**A**) ruminants and (**B**) hindgut fermenting species to extract energy (expressed relative to daily maintenance requirements) from abundant foods.

more, and this outweighs the consequent reduction in digestive efficiency.

2 Large animals can obtain more energy from a given food than small animals, because their longer retention times allow more efficient digestion and this permits greater intake relative to $W^{0.73}$.

3 The scaling of potential energy intake with body weight averages $W^{0.85}$ (see Illius & Gordon 1991, 1992), producing a marked selection pressure for the evolution of large body size under conditions of abundant food.

4 Comparing foods with equivalent potential digestibility but differing slightly in composition and digestion kinetics shows up large differences in potential energy yield to a given animal. It also shows differences in the relative value of foods to ruminants and hindgut fermenters. In effect, digestive constraints are a complex function of the effects of body size, anatomy and food properties. They are certainly not a simple function of the food alone, equally applicable to any animal.

It is clear that the quality of grass which is adequate for the grazers can only provide about 75% of the requirements of the browsers, demonstrating the importance of body size for diet selection. However, more highly digestible grass, for instance fresh growth following rains, would be sufficient to sustain the smallest animals, and this is accompanied by seasonal shifts in the diets of these animals (see also Gordon & Illius 1988). The quality of forbs which would maintain the gazelle would provide 1.6 times maintenance if eaten by the buffalo (1.3 for the wildebeest). Why then are forbs and browse not selected by the two larger animals? The reason may be that forbs have low biomass compared with grasses, and occupy the base of the vegetation profile. Browse is usually more widely dispersed than grasses, generally as single items (leaves, fruit) on perennial woody tissue. Grazing animals have a dental morphology which allows a broad and flattened incisor arcade to be presented to the vegetation, but prevents the accurate selection of individual plant parts, while browsers have narrower and more pointed incisor arcades, capable of greater selectivity (Gordon & Illius 1988). Even if the large grazers could select such food items, they could not contribute much to energy intake because of their small size. If it is assumed that a browser could take up to 15 000 food items in a day (Gordon 1986), then the food items of a sufficient size to meet the smallest animal's energy needs would only provide a fraction of the needs of larger animals. This is shown in Table 8.2, where the minimum item sizes of the forbs and browse adequate for gazelle and impala are inadequate for larger animals

Table 8.2 Daily energy intake, expressed in multiples of maintenance energy needs, from small items of food

	W (kg)	Forbs item size* (mg)	Energy intake	Browse item size* (mg)	Energy intake
Thomson's gazelle	20	3.6	1.00	8.5	0.90†
Impala	60	3.6	0.45	8.5	1.00
Wildebeest	180	3.6	0.21	8.5	0.46
Buffalo	540	3.6	0.09	8.5	0.20

* Chosen to suit gazelle or impala.
† Limited by digestive capacity.

if all animals were to take the same number of them. This is a reasonable assumption given that large animals cannot feed much more rapidly or for longer than small animals (Gordon 1986). So the distribution of food item sizes and potential digestibility determine broadly the diets that animals of different body size must select. A similar explanation probably applies to the question of why equids do not use browse and forbs: presumably they occur at too low a density to support the high intakes required by equids (see Janis, Gordon & Illius, in press).

What are the implications of body size for intake rate from grazed swards as compared with intake rate of browse food items of finite size? The following results quantify the disadvantages of large body size under conditions of resource depletion, where low biomass limits intake rate. If the guild of animals were to graze from short swards, maintained in a highly digestible and vegetative state by grazing pressure (McNaughton 1984; Illius, Wood-Gush & Eddison 1987) then it has been shown that the small animals will have an advantage over large animals (Illius & Gordon 1987). This was demonstrated by a model of grazing processes which describes how dental morphology and vegetation structure interact to determine the weight and composition of plant material which can be removed in a single bite. Comparing animals taking 30 000 bites a day (close to the maximum) shows how body size and variation in vegetation characteristics such as height, biomass concentration and vertical distribution of live and dead plant parts affects the functional response (daily energy intake rate relative to requirements). The model incorporates the relationships between animal size, digestive efficiency and physical constraints on intake examined above. The model was used to predict energy intake from a medium density sward with a majority of highly digestible components, and with 20% dead material at the base of the sward profile. The results show that short swards impose greater limitations on food intake by large animals than smaller animals (Fig.8.3a).

On swards with a surface height of 30 mm, the buffalo is severely restricted (half maintenance) while the gazelle can obtain maintenance, and the other two animals can get 0.9 of maintenance. Above this sward height, the comparison is more complex, because differences in dental morphology between browsers and grazers begin to have an effect, and the gazelle runs into digestive capacity constraints. On sparse swards with a large proportion of dead material (Fig. 8.3b), the combined effects of incisor morphology constraints and digestive capacity constraints cause a reversal of the scaling of intake with body size from $W^{0.36}$ (on very short swards) to $W^{0.85}$ (on taller swards). The implication of this for foraging choices is that the relative value of two alternative food patches, for example a tall mature patch of herbage and a short vegetative sward, depends not merely on the foods' properties but on the size of the animal confronted by that choice. Size-related constraints on food intake rate on low-biomass vegetation can explain a number of ecological phenomena of grazing ungulate communities, such as grazing succession, facilitation and the segregation of habitat use in sexually dimorphic species (see Illius & Gordon 1987). These are clearly the result of individuals' foraging decisions, and it is argued that these decisions are made under constraints which are set according to body size. Preliminary evidence from a controlled test supports this hypothesis: intake rate on short swards was reduced in large animals to a greater extent than in small animals, and scaled with $W^{0.42}$ (Illius, Gordon & Milne, unpublished).

The combined effects of these constraints on niche selection can be generalized across the distribution of vegetation in two dimensions of quality and quantity (Fig. 8.4, *see* page 167). Each body size is associated with two zones: an exclusive zone in which either the quantity or quality of food is inadequate for other species (niche separation); and a common zone which can also be utilized by other species (niche overlap). Empirical evidence to support this formulation comes from the finding that the rumen fermentation rates of diets selected by a range of African bovids scale with $W^{-0.23}$ (Gordon & Illius, submitted). This shows that the quality of the diet selected is lower in large animals than in smaller ones, in agreement with predictions (Fig. 8.4). Large animals thus have poorer quality diets than small animals because the supply of high-quality foods

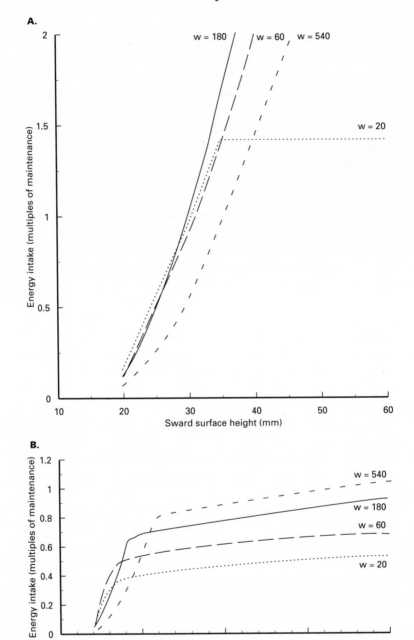

Fig. 8.3 Sward height effects on daily energy intake of animals differing in weight on (**A**) vegetative and (**B**) senescent swards.

is inadequate to meet their large requirements (see Grant *et al.* 1985, discussed below, for an experimental demonstration of this).

The comparison of ruminants and hindgut fermenters (Fig. 8.2a,b) suggests that there is unlikely to be competition in similar food quality niches between species of similar size but with different digestive systems, or between small ruminants and large hindgut fermenters. However, smaller hindgut fermenters, which can consume a medium quality diet, are likely to compete directly with large ruminant species, because the former can tolerate lower food availability. For example, the 200 kg plains zebra probably competes directly with the 540 kg African buffalo for the same feeding niche (Hansen *et al.* 1975), to a greater extent than it competes with the similarly-sized wildebeest (Gwynne & Bell 1968; Owaga 1975). During the dry season, zebra and buffalo diets have been found to have the same diet composition (Gwynne & Bell 1968; Sinclair and Gwynne 1972) and to use vegetation of similar biomass concentration and structure (McNaughton 1985).

The modelling exercise suggested that the interaction of body size and food properties is such that, given a seasonal cycle of food quality and quantity, the scaling of energy intake ranges from $W^{0.85}$ to $W^{0.36}$.

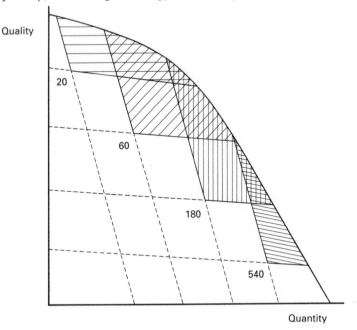

Fig. 8.4 Feeding niches defined by food quality and quantity constraints for animals of different body size.

The implication is that the evolution of body size has taken place under counterbalancing selection pressures during seasons of abundance and depletion, modified by the relative duration of those seasons and by the size of the body reserves. It follows that animals have evolved to exploit vegetation within subsets of the quantity–quality range. One would therefore expect animals to be able to recognize their feeding niche as the quantity–quality space in which diet selection is least constrained. On this view, size-related constraints underlie a perceptual construct – the recognized feeding niche.

FORAGING STRATEGIES AND TACTICS

Having defined the subset of the vegetation that constitutes the feasible set of a herbivore's foods, the foraging strategies which are likely to enhance the animal's fitness can now be discussed. These are narrowly defined here in terms of the fulfilment of its nutrient requirements. The maximization of energy intake rate often serves as convenient proxy for fitness maximization, and rate maximization is assumed to apply in normative models of mammalian herbivore foraging (Belovsky 1978; Owen-Smith & Novellie 1982; Ungar & Noy-Meir 1988). But there are a number of grounds for questioning whether energy maximization applies to vertebrate herbivores (Crawley 1983). For example, it has been argued that, for herbivores, maximizing the rate of energy intake may be less important in diet selection than obtaining a balanced diet and avoiding toxins and other anti-nutritional plant compounds (Westoby 1974, 1978). However the evidence that herbivores select nutrients to balance their diets is weak, and particular dietary or anti-nutritional factors can be incorporated into rate-maximizing models by specifying constraints (Stephens & Krebs 1986).

The complexity of vegetation structure and composition again lies at the centre of the problem, with variability in nutrient content hypothetically imposing the need for mixed diets in order to balance nutrient intake. Moreover, the structural variability of vegetation implies that the need for sampling and the problem of discrimination will constrain the achievement of a diet offering maximum nutrient intake rate. That said, the major source of variation in the nutrient content of vegetation is the presence of lignocellulose in maturing plant tissues, the composition of the digestible fraction being much less variable. Thus both energy maximization and nutrient maximization are foraging strategies which are likely to be achieved by tactics which maximize the intake rate of digestible plant tissues. Therefore rate maximization seems the most likely explanation of herbivore foraging behaviour, subject to the avoidance of harmful chemical compounds and constrained by the inherently difficult task of discrimination between foods on the basis of their ultimate nutrient yield.

There is some evidence that herbivores do actually select diets which maximize their rate of intake. The major variables governing intake rate are the structural properties of the vegetation such as height, density and the vertical distribution of biomass (Black & Kenney 1984; Hodgson 1985; Burlison, Hodgson & Illius 1991). Tall vegetation with high biomass density allows higher rates of intake than short vegetation, and provided each vegetation type is of similar digestibility, the taller vegetation is preferentially grazed (Arnold 1987; Bazely 1990). However, the selection of shorter high-quality vegetation may occur when the taller vegetation type is of lower digestibility, suggesting that the choices herbivores make result in increased nutrient intake rate by comparison with the average available, and not just increased bulk intake. This is illustrated by the diets selected by sheep and cattle grazing hill grasslands composed of a mosaic of *Nardus* tussocks growing in a short background sward of more highly-digestible grasses (principally *Festuca* and *Deschampsia*). Grant *et al.* (1985) showed that both animal species avoid *Nardus*, but as the preferred inter-tussock sward becomes progressively depleted, the proportion of *Nardus* in the cattle diet rises from 0.1 to 0.5, while that in the sheep diet rises from only 0.05 to 0.09. The example shows the greater tendency of the larger species to be prepared to eat more fibrous food, coupled with greater sensitivity to low biomass swards. It also illustrates a common feature of ungulate diets: that they tend to be composed of not only the single most 'profitable' component, but also a certain proportion of the less-preferred component (e.g. *Nardus*). Sheep and cattle diets merely differ in the proportions of these components, arising from the sheep's higher selectivity and the tolerance by cattle of a poorer-quality diet (Grant *et al.* 1985). It is presumed, but has not actually been demonstrated, that differences in the size and morphology of each species lead it to use different grazing tactics in pursuit of intake rate maximization.

More theoretically-based approaches have also supported energy maximization as the major diet selection strategy of herbivores. Belovsky (1986) concluded that virtually all generalist herbivores are energy maximizers rather than time minimizers: that is, they aim for maximum energy intake rather than for fulfilment of a limited requirement in minimum time. Breeding males may be the exception, he argues, because of their need to allocate time to reproductive activities. Belovsky uses linear programming to determine the effect of a range of constraints on the intake of broad classes of vegetation, but the technique cannot explain the plant species composition of the diet (Belovsky 1984). Vivås *et al.* (1991) also made a successful attempt at explaining diet choice by assuming that energy maximization and digestive constraints apply in moose foraging on birch twigs. They predicted optimal twig size from the trade-off between increasing biomass intake rate and the declining digestibility of the biomass as the size of twigs chosen was increased, and this coincided quite closely with the size of twigs moose were observed to take.

Browsers such as moose exploit discrete food items (twigs and leaves) which are patchily distributed (on trees) and accordingly their foraging has been considered as appropriate for description by optimal patch use models (Lundberg & Aström 1990). Of course it would be surprising if animals did not respond to the depletion of a patch of vegetation by leaving it for another, but the question is whether patch residence time by herbivores can be predicted from the gain function and travel time according to the classical concept of marginal value (Charnov 1976). Although the requisite decelerating gain functions were established for moose by Aström, Lundberg and Danell (1990), the functions barely differed from linearity and had shallow slopes, giving a wide range of predicted optimal patch residence times. Moose were observed to have very variable residence times for any given tree size, thus giving rather unconvincing support for marginal value predictions. Sheep have been observed to forage in a manner consistent with the marginal value theorem when presented with a few clearly-distinguishable patch types (Bazely 1990), although it is hard to imagine that natural vegetation communities present such obvious choices.

INFORMATION CONSTRAINTS

Early models of optimal foraging can be criticized for requiring the rather stringent assumption that animals have complete information about the relevant parameters of the environment (i.e. prey or patch characteristics, handling or travel times) as discussed by Stephens and Krebs (1986). The problem of incomplete information that besets any animal living in a stochastic environment is likely to apply to herbivores, since natural vegetation is highly variable over space and time and does not occur in discrete patches with easily recognizable intake characteristics. Local variation in sward surface height may present an obvious cue to the animal, but intake rate is also likely to be influenced by the more cryptic structural properties of the base of the sward, which affect bite depth and diet composition (e.g. Barthram 1981; Grant *et al.* 1985). The existence of stochastic variation in sward structure, which is not closely related to the more visible surface properties of the vegetation, implies that each bite or item taken has a profitability that is unlikely to be apparent prior to consumption, and yields an instantaneous intake rate that may vary widely over short periods of time. For herbivores it is inappropriate to talk of specific and recognizable food items, because vegetation is only partially consumed and the extent of partial consumption has a stochastic element. This suggests that the herbivore's problem of incomplete information is partly one of recognizing patch quality. The animal is therefore faced with ambiguity about the value of the items at hand and about the items in alternative patches. With limited abilities at prior recognition of item type, it must sample the vegetation while deciding where to graze.

Only by understanding the problems that the structural and spatial heterogeneity of vegetation pose for the animal can one hope to deduce what decision rules are likely to have evolved as effective foraging tactics. It would be fair to claim that little enough is known of these problems for the investigation of herbivore decision-making still to be in the descriptive phase, and that it would be premature to attempt a formulation of theoretically optimal behaviour. Several experiments have observed the diet selection of herbivores faced with making foraging choices between clearly defined alternatives, and we review some the results below. The question to pose is: what does the animal's performance in these tasks indicate about the information constraints on maximizing intake?

In an experiment to determine the effect of sward height on patch selection, Clark *et al.* (unpublished) observed groups of cattle, sheep and goats grazing large plots composed of perennial ryegrass monoculture. Each half of each plot had been mown to a predetermined height to give a range of paired height comparisons: 2 cm vs. 4, 2 vs. 8, 4 vs. 8, 4 vs. 16, 8 vs. 16, 8 vs. 32 and 16 vs. 32. These differences were quite clear to the human eye, with an obvious boundary across the middle of each plot. Bite weights increased 3–4 fold over the range of increasing sward height, but bite rate declined by 50%, and thus intake rate increased by a factor of 2.

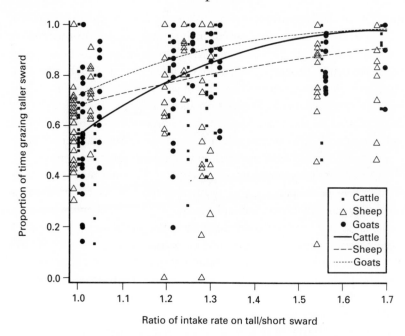

Fig. 8.5 Proportion of time grazing the taller of two patches in relation to the ratio of the intake rates on tall and short patches.

The taller swards generally were strongly preferred by all species. The results are plotted in Fig. 8.5, showing data and regressions of the arcsin proportion of time grazing the taller patch on the ratio of the intake rates on tall and short patches. As the intake rate ratio (IRR) increased, each species spent a greater proportion of time grazing the taller sward. The purpose of describing selection (time grazing the taller sward) as a function of IRR was to generalize across the choices offered (although IRR does not necessarily relate linearly to animals' perceived superiority of the taller swards). Given that animals prefer swards with higher intake rates, the expected degrees of selection for extremes of IRR are, respectively, 0.5 and 1 for IRR = 1 and ∞. The shape of the response curve between these extremes can then be interpreted as the sensitivity of an animal to choices where IRR > 1.

The results of the regression analysis show some differences between animal species. Cattle were more sensitive to IRR than sheep, showing 0.9 selection of the taller sward at IRR = 1.35. Sheep showed the lowest sensitivity to IRR, with 0.9 selection of the taller sward being predicted from the regression (by extrapolation) at IRR = 2. Goats showed the greatest sensitivity, grazing for 0.9 of the time from swards with IRR = 1.3. The goats were found to be taking shallower bites from the sward surface than were the other species, especially on the taller swards. The regressions imply that the animals were fairly sensitive to height or intake rate differences, but it can be seen from the scatter of the data points (Fig. 8.5) that there is considerable unexplained variation in patch selection (after transforming the data and taking account of the experimental design $r^2 = 0.634$). This can partly be explained by the animals' movement patterns, which contain a random element. During grazing, the animals' movements about the plot often (every 3–6 min) took them back onto the shorter sward. Since they only stayed there briefly, and given that this was a controlled experiment, it can be concluded that they responded to information gained on the shorter sward, and therefore their apparently random movements also served a sampling function. But can *purposive* sampling to gain information be construed from movement patterns during foraging? The more obvious function of movement should be considered first.

Movement patterns can be rationalized in terms of the conflict between the need to exploit depleting food patches and the need to move on to search for, sample and exploit new patches. The balance between stationary exploitation and moving-searching will thus depend on the habitat's richness and the patchiness and dispersal of food. This appears to boil down to the classical problem of foraging on patchy resources addressed by the marginal value theorem of Charnov (1976), but we should be cautious about the usefulness of this conclusion as it stands. The marginal value theorem provides the theoretical solution to a problem defined by known patch characteristics, but as Stephens and Krebs (1986) make clear, it does not provide a decision-rule to be implemented by animals exploiting patches in an uncertain and stochastic environment. Nevertheless, we can expect animals to have evolved effective and robust tactics for solving the foraging problems they normally encounter, and for these to approximate to optimal solutions. Movements during grazing seems to be a natural part of the herbivore's foraging tactics: animals move forwards as they graze, and never stay in one place for very long, especially if they chance on an inferior patch (as in Clark's experiment). Moving backwards and forwards between patches is not apparently related to patch depression, since it occurs even on large patches where the depletion by grazing is trivial. The reason why herbivores leave patches 'early' may be their need to gain information about the habitat: in other words, patch assessment is, besides patch depression, 'another reason to move on' (Stephens & Krebs 1986). This would suggest that movement, patch exploitation and patch assessment may be virtually concurrent processes. Sampling by herbivores as they move between patches may explain their 'partial preferences' and frequency-dependent selection, as Illius, Wood-Gush and Eddison (1987) observed in cattle grazing heterogeneous swards. Cattle showed a marked preference for short but more digestible patches over intermediate patches and over tall patches of lower digestibility, but diet selection was also a function of the frequency with which each patch type was encountered.

Assuming that the herbivore's movement patterns partly serve a sampling function, the fact that the animals in Clark's experiment frequently left the longer sward suggests that they needed repeatedly to sample the shorter sward, and to do so by consumption (i.e. by actually grazing from it), rather than by relying on sight. Similar observations were made by Jarman (pers. comm.) on feeding preferences in impala. Jarman offered impala a choice between pairs of familiar species of *Acacia* (a genus of browse plants varying widely in thorniness and content of secondary compounds), one of which was distinctly preferred. The impala ate a small amount of each before rejecting one (especially the more thorny species) and feeding on the other. But every few minutes they returned to the previously-rejected species and took a couple of bites before returning to the preferred species: merely looking at the rejected species was not enough. Given the strength of preference, why did the impala keep returning to the alternative species? It seems reasonable to speculate that this characteristic behaviour is something to do with the way and the timescale over which herbivores gain and process information. For example, the impala might have been cautiously sampling toxins with a delayed effect (see Provenza *et al.* 1991), and indeed all sampling may reflect this caution. But if the information about patch quality is immediate (e.g. an aversive taste, the presence of thorns, or distinct difference in sward height) we might surely expect herbivores to retain such information *provided* it is normally worth doing so, and therefore not to require to sample repeatedly. Information would not be worth retaining if it was of trivial importance for fitness (e.g. in the choice between two very similar non-toxic foods) or had low predictive value. Arguably, herbivores have evolved to exploit heterogeneous environments which normally supply food at such a low rate that animals have to keep moving on during the course of foraging. In that case, it would be of little use to an animal to retain for long the information about food patches that it only visits briefly. Thus limited retention of information and repeated sampling could reflect the low value of information in a rapidly changing world (Stephens 1987), and the correspondingly low selection pressure on the ability to remember the precise properties of non-toxic foods (but *see* Chapter 5 for further discussion on learning).

In this section it has been argued that the foraging tactics of herbivores often include movement patterns which allow the animal to gain information and respond. Thus the partial preferences which herbivores exhibit when given controlled experimental choices between food patches could arise in part from their need to gather information. There are also questions raised about what it is worth a herbivore knowing about its nutritional environment.

PERCEPTION AND INFORMATION-GATHERING

Clark's experiment (*see* page 171) estimated molar responses to vegetation monocultures intended to differ only in height. Illius, Clark and Hodgson (1992) carried out an experiment designed to test how sheep responded to a more complex choice, and allow short-term responses to be observed more closely than is possible in the field. This was achieved by using pairs of swards differing in height and grass/clover content, and grown in seed trays. Different mixtures of grass and clover were obtained by varying the seed ratio from 0% to 100% clover in 10% intervals. Sheep were given a choice between swards covering a wide range of contrasts in clover content, and these contrasts were either pre-trimmed to produce swards of nearly equal surface height or differing by 4 cm in height. The sheep grazed readily from the swards, switching between alternate swards after bouts of 20–30 bites. Regression analysis showed that patch selection was influenced by sward height and by clover content, and their effects were additive. Tall patches were preferred, and intermediate contents of clover were selected in preference to low or high contents. The additive effect of combined height and clover differences is shown in Fig. 8.6a. The fitted response surface shows the proportion of bites (y-axis) taken from a sward varying in clover content (x-axis) and the surface height (z-axis) when the alternative sward has 50% clover and a height of 10 cm. As with Clark's data, the regressions left considerable unexplained variation ($r^2 = 0.45$).

Were the sheep making choices which tended to maximize their intake rates? Intake rate was affected by both clover content and sward height (Fig. 8.6b,c) and the intermediate clover contents and tall swards on which intake rate was highest coincided with the preferences shown.

This coincidence of the intake-rate response and the selection response to changing clover content is striking, but superficial. Neither the proportion of time spent grazing a sward nor the proportion of intake from the sward was closely related to the **actual** differences in intake rate between the alternatives (as distinct from the **overall** effect of clover content and sward height on intake rate shown in Fig. 8.6b,c). Intake rate differences only accounted for about 12% of the variation accounted for by height differences. In other words, height was a good predictor of selection, but intake-rate differences were not. Nevertheless, selecting taller swards would, on average, result in higher intake rates than would unselective grazing (Fig. 8.6c), so it is plausible that sward height could be used as a cue to increase intake rate in the long run.

One reason why intake rate itself appeared to be of limited use in short-term decision-making is that it was affected not only by the composition of the patch being grazed but also by the composition of the previously-grazed patch. When sheep switched from one grazing site to another it appeared that there was some temporary carry-over of some of the grazing style they had developed on the previous sward. Intake rate was maximized at higher clover contents as the clover content of the alternative sward increased. For example, intake rate was maximized on swards with 30, 45 or 60% clover where the alternative sward content was 0, 50 or 100% respectively (Fig. 8.6b). Carry-over effects of the clover content of the alternative sward accounted for about a third of the total variance explained by all clover terms. In the short term, intake rate does not apparently provide a pure signal of patch quality and this must increase the discrimination difficulties on heterogeneous vegetation. Given longer runs of bites on homogeneous patches this carry-over effect would be minimized and thus intake rate would provide a purer signal of patch profitability. Similar observations were made in the signal detection analysis of Getty and Krebs (1985), who found that great tits had lagged responses to changes in prey frequency, showing the carry-over of habituation to previously experienced prey frequencies. Both studies show that foraging decisions are influenced not only by incident prey conditions, but also by expectations or techniques which reflect past experience.

Can this experiment give any idea what sort of information sheep actually do use in short-term decision-making and how they gather it? Decisions based on prior recognition of patch types can only have had a limited effect on diet selection, because after making an initial choice, animals switched frequently between patches, indicating that some form of decision-making occurred **during** grazing. Sheep tended to choose the

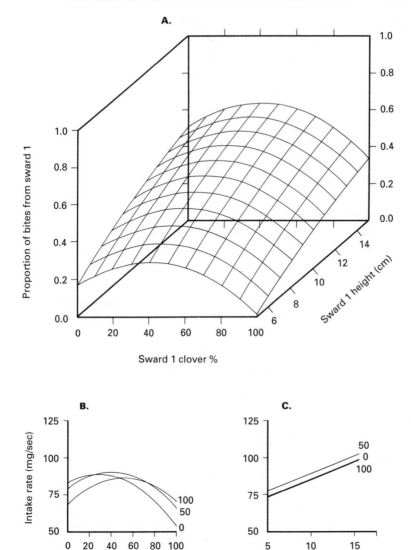

Fig. 8.6 (**A**) The proportion of bites taken from a sward varying in clover content and surface height when the alternative sward has 50% clover and a height of 10 cm. (**B**) Response of intake rate to variation in clover % at 10 cm sward surface height, and (**C**) to variation in sward surface height at 50% clover. The effects of alternative sward composition at 0, 50 and 100% clover are also shown.

taller sward first, provided that the height difference was sufficiently large. But in the trials where height differences between patches were minimized, the small remaining differences in height had no effect on which sward was chosen to be grazed first, yet the sheep still went on to exploit taller patches to a greater extent than short patches. Furthermore, the clover contents of each alternative sward did not affect initial patch choice, yet clover content influenced the proportion of bites ultimately taken from a patch. This suggests that sheep make limited use of pre-consumption cues and may not detect by sight the patch characteristics that they will respond to during grazing. Positive reinforcement is evidently gained from preferred swards during a biting bout, implying that 'handling' cues (such as ease and effort of handling) are used, as well as visual cues.

Further evidence of the sort of information to which sheep respond was obtained by analysing whether they responded to local variations in surface clover content within each sward. It has already been shown that they did respond to differences in the *average* clover content of patches (Fig. 8.6a). The analysis showed that within short bouts of biting from each patch the sheep did not take the opportunity to exploit heterogeneity in surface clover content. This suggests that neither pre-consumption cues nor the individual bite provides sufficient information to have much influence on selection, and that each bite is just one of a number of samples. On turning to a sward the sheep grazed rapidly (c. 1 bite sec^{-1}) from a confined area before switching to another location in the same sward or in the alternative sward. It therefore seems likely that information gained from each bite is accumulated over successive bites, and that switching decisions are based on short periods of grazing rather than on individual bites. From this it would be expected that termination of a biting bout would be delayed on preferred patches. Indeed, bout length was found to be longer on taller swards and at intermediate clover levels.

The results can be summarized as follows. Sheep did not respond to patch differences in intake rate, although height differences were detected during short bouts of biting. Clover content was detected to the point that allowed selection to be exercised over a succession of bouts, but the sheep did not direct successive bites to local variations in clover content. These results imply that the discrimination problem posed by this experiment was close to the lower limit of the sheep's spatial and temporal scales of perception and decision-making. Yet the task, which involved discrimination between two continuously variable mixtures rather than between different foods, was no more complex than a grazer might face with every step over natural vegetation.

CONCLUSIONS

Mammalian herbivores are constrained in the diets they can select by variables associated with their body size and perceptual faculties. Body size determines the efficiency with which a given food can be consumed and utilized by setting quantity and quality constraints, and these together define the limits of the feeding niche. Within these limits, patch and food item selection operate. It can be argued that the wide dispersal of acceptable food items within structurally variable vegetation has led to the evolution of a style of feeding which involves frequent movement, both as vegetation becomes depleted and to gain information. Sheep seem to act on information accumulated over short bouts of grazing, and exercise preference by deciding whether to extend or terminate these bouts. On this view, the duration of the bout is the decision variable, and the diet selected is a function of the number and length of bouts on each patch type. Although the value of information in a world which changes frequently in time and space is low, the cost of gaining the information is probably low as well. This is because when environmental variance (the contrast between the quality of patches) is low, the reduction in intake rate from sampling a poor patch is also low; while high environmental variance makes inferior patches easier to detect and so requires a smaller and hence less costly sample. In either case, the modification of bout length to that required to detect an inferior patch serves to minimize patch assessment costs, subject to the animal's perceptual acuity. The herbivore's imperfect perception of its environment means that it has to gain information while feeding. The overlap between these two processes could explain the characteristic partial preferences herbivores show as they turn from one food source to another, and implies that diet optimization cannot occur on a fine-grained scale of time and space.

REFERENCES

Arnold G.E. (1981) Grazing behaviour. In *Grazing Animals* (ed. by F. Morley). Elsevier, Amsterdam.

Arnold G.E. (1987) Influence of the biomass, botanical composition and sward height of annual pastures on foraging behaviour by sheep. *J. Appl. Ecol.* **24**, 759–72.

Aström M., Lundberg P. & Danell K. (1990) Partial prey consumption by browsers: trees as patches. *J. Anim. Ecol.* **59**, 287–300.

Barthram G.T. (1981) Sward structure and the depth of the grazed horizon. *Grass and Forage Science* **36**, 130–1.

Bazely D.R. (1990) Rules and cues used by sheep foraging in monocultures. In *Behavioral Mechanisms of Food Selection* (ed. by R.N. Hughes) *NATO ASI series, vol. G 20.* Springer Verlag, Berlin.

Bell R.H.V. (1970) The use of the herb layer by grazing ungulates in the Serengeti. In *Animal Populations in Relation to their Food Resources* (ed. by A. Watson), pp. 111–23. Blackwell Scientific Publications, Oxford.

Belovsky G.E. (1978) Diet optimization in a generalist herbivore: the moose. *Theor. Popul. Biol.* 14, 105–34.

Belovsky G.E. (1984) Herbivore optimal foraging: a comparative test of three models. *Amer. Natur.* 124, 97–115.

Belovsky G.E. (1986) Optimal foraging and community structure: implications for a guild of generalist grassland herbivores. *Oecologia (Berlin)* 70, 35–52.

Black J.L. & Kenney P.A. (1984) Factors affecting diet selection by sheep. II. Height and density of pasture. *Aust J. Agric. Res.* 35, 565–78.

Burlison A.J., Hodgson J. and Illius A.W. (1991) Sward canopy structure and the bite dimensions and bite weight of grazing sheep. *Grass For. Sci.* 46, 29–38.

Charnov E.L. (1976) Optimal foraging: the marginal value theorem. *Theor. Popul. Biol.* 9, 129–36.

Crawley M.J. (1983) *Herbivory.* Blackwell, Oxford.

Demment M.W. & Van Soest P.J. (1985) A nutritional explanation for body-size patterns of ruminant and nonruminant herbivores. *Am. Nat.* 125, 641–72.

Geist V. (1974) On the relationship of ecology and behaviour in the evolution of ungulates: theoretical considerations. In *The Behaviour of Ungulates and its Relation to Management* (ed. by V. Geist & F.R. Walther), pp. 235–46. IUCN, New Series No 24.

Getty T. & Krebs J.R. (1985) Lagging partial preferences for cryptic prey: a signal detection analysis of great tit foraging. *Am. Nat.* 125, 39–60.

Gordon I.J. (1986) *The feeding strategies of ungulates on a Scottish moorland.* PhD Thesis, University of Cambridge.

Gordon I.J. & Illius A.W. (1988) Incisor arcade structure and diet selection in ruminants. *Funct. Ecol.* 2, 15–22.

Grant S.A., Suckling D.E., Smith H.K., Torvel L., Forbes T.D.A. & Hodgson J. (1985) Comparative studies of diet selection by sheep and cattle: the hill grasslands. *J. Ecol.* 73, 987–1004.

Gwynne M.D. & Bell R.H.V. (1968) Selection of vegetation components by grazing ungulates in the Serengeti National Park. *Nature, Lond.* 220, 390–3.

Hansen R.M., Mugambi M.M. & Bauni S.M. (1975) Diets and trophic ranking of ungulates of the northern Serengeti. *J. Wildl. Manage.* 49, 823–9.

Hobbs N.T., Baker D.L. & Gill R.B. (1983) Comparative nutritional ecology of montane ungulates during winter. *J. Wildl. Manage.* 47, 1–16.

Hodgson J. (1985) The control of herbage intake in the grazing ruminant. *Proc. Nutr. Soc.* 44, 339–46.

Illius A.W., Clark D.A. & Hodgson J. (1992) Discrimination and patch choice by sheep grazing grass-clover swards. *J. Anim. Ecol.* 61, 183–94.

Illius A.W. & Gordon I.J. (1987) The allometry of food intake in grazing ruminants. *J. Anim. Ecol.* 56, 989–99.

Illius A.W. & Gordon I.J. (1991) Prediction of intake and digestion in ruminants by a model of rumen kinetics integrating animal size and plant characteristics. *J. Agric. Sci., Camb.* 116, 145–57.

Illius A.W. & Gordon I.J. (1992) Modelling the nutritional ecology of ungulate herbivores: evolution of body size and competitive interactions. *Oecologia (Berlin)* 89, 428–34.

Illius A.W., Wood-Gush D.G.M. & Eddison J.C. (1987) A study of the foraging behaviour of cattle grazing patchy swards. *Biol. Behav.* **12**, 33–44.

Jarman P.J. (1974) The social organization of antelope in relation of their ecology. *Behav.* **48**, 215–66.

Jarman P.J. & Sinclair A.R.E. (1979) Feeding strategies and the pattern of resource partitioning in ungulates. In *Serengeti: Dynamics of an Ecosystem* (ed. by A.R.E. Sinclair & M. Norton–Griffiths), pp. 130–63. University of Chicago Press, Chicago.

Lundberg P. & Aström M. (1990) Functional response of optimally foraging herbivores. *J. Theor. Biol.* **144**, 367–77.

McNaugton S.J. (1984) Grazing lawns: animals in herds, plant form and coevolution. *Am. Nat.* **124**, 863–86.

McNaughton S.J. (1985) Ecology of a grazing ecosystem: the Serengeti. *Ecol. Monogr.* **55**, 259–94.

Owaga M.L. (1975) The feeding ecology of wildebeest and zebra in Athi-Kaputei plains. *E. Afr. Wildl. J.* **13**, 375–83.

Owen-Smith N. & Novellie P. (1982) What should a clever ungulate eat? *Am. Nat.* **119**, 151–78.

Penry D.L. & Jumars P.A. (1987) Modeling animal guts as chemical reactors. *Am. Nat.* **129**, 69–96.

Peters R.H. (1980) *The Ecological Implications of Body Size.* Cambridge University Press, Cambridge.

Poppi D.P., Minson D.J. & Ternouth J.H. (1981) Studies of cattle and sheep eating leaf and stem fractions of grasses. III. The retention time in the rumen of large feed particles. *Aust. J. Agric. Res.* **32**, 123–37.

Provenza F.D., Burrit E.A., Clausen T.P., Bryant J.P., Reichardt P.B. & Distel R.A. (1991) Conditioned flavor aversion: a mechanism for goats to avoid condensed tannins in blackbrush. *Am. Nat.* **136**, 810–28.

Short H.L., Blair R.M. & Segelquist C.A. (1974) Fiber composition and forage digestibility by small ruminants *J. Wildl. Manage.* **38**, 197–209.

Sibly R.M. (1981) Strategies of digestion and defecation. In *Physiological Ecology* (ed. by C.R. Townsend & P. Calow). Blackwell Scientific Publications, Oxford.

Sinclair A.R.E. & Gwynne M.D. (1972) Food selection and competition in the East African buffalo (*Syncerus caffer* Sparrman). *E. Afr. Wildl. J.* **10**, 77–89.

Stephens D.W. & Krebs J.R. (1986) *Foraging Theory.* Princeton University Press, N.J.

Stephens D.W. (1987) On economically tracking a variable environment. *Theor. Popul. Biol.* **32**, 15–25.

Taylor StC.S. (1980) Genetic size scaling rules in animal growth. *Anim. Prod.* **30**, 161–5.

Ungar E.D. & Noy-Meir I. (1988) Herbage intake in relation to availability and sward structure: grazing processes and optimal foraging. *J. Appl. Ecol.* **25**, 1045–62.

Vivås H.J., Sæther B.-E. & Andersen R. (1991) Optimal twig-size selection of a generalist herbivore, the moose *Alces alces*: implications for plant–herbivore interactions. *J. Anim. Ecol.* **60**, 395–408.

Westoby M. (1974) An analysis of diet selection by large generalist herbivores. *Am. Nat.* **108**, 290–304.

Westoby M. (1978) What are the biological bases of varied diets? *Am. Nat.* **112**, 627–31.

9: Effects of Ecological Interactions on Forager Diets: Competition, Predation Risk, Parasitism and Prey Behaviour

ANDREW SIH

INTRODUCTION

Imagine a fish in a pond, foraging on aquatic insects in shallow-water vegetation, on zooplankton in deeper, open water, and on worms in the muddy bottom. It must choose a favourite foraging habitat, and of the foods in that habitat, it must choose to attack some and ignore others. The outcome of this set of decisions is its diet, which has major effects on growth, survival and future production of fish fry. As the old saying goes 'You are what you eat'.

The fish's diet is of interest not just to the fish itself, but also to ecologists. Behavioural ecologists are inherently interested in understanding the adaptive significance of dietary patterns (Pyke 1984; Stephens & Krebs 1986; Hughes 1990). Community ecologists are interested in diet because what is eaten can affect interactions between the organism and its competitors (Abrams 1983; Schoener 1986) and can have a major influence on its impact on the prey community (Sih *et al.* 1985).

The dominant conceptual approach for understanding diets is optimal diet theory (ODT). Classical ODT predicts diets under the assumption that foragers attempt to maximize their net rate of energy intake. This approach, however, has many limitations. Energy might not be the most important dietary component: other nutrients such as proteins, minerals and vitamins, and toxins may be of interest, as may other fitness factors (e.g. predation risk while feeding) that can depend on diet choice. Classical ODT assumes, in essence, that an organism's foods are nutritionally complete (contain all necessary nutrients), and that the organism has no aggressors, predators or parasites. While this may be true for some foragers, particularly in laboratory conditions, this is clearly not true for many, if not most foragers in nature. One focus of this chapter is on the effects of other organisms on diets. How should (based on adaptive considerations) and how do competitors or predators influence diets?

ODT makes predictions about active predator choice (the decision to attack some prey and reject others). This, however, is only one component of preference. Preference is defined in the ecological literature as a tendency to eat more of a prey type than would be expected based on its abundance in the habitat (Chesson 1983). What is eaten is determined by what is encountered, the choice of prey to attack, what can be caught and what can be consumed. Preference can be generated by variations among prey in any of these stages. Active predator choice can determine preferences, but ecologically significant preferences can occur without any active predator choice. That is, a prey type might be preferred simply because it is seen more often (e.g. it does not hide), or it is caught more easily than other prey (e.g. it does not have effective means of escaping). Note that preference thus depends on both prey and predator behaviour. A second focus of this chapter is on the effects of prey behaviour on forager diets.

Thus, this chapter: (i) introduces a conceptual framework for addressing how interactions among individuals might influence preference by influencing the components of the predator–prey interaction; (ii) briefly reviews the predictions of optimal diet theory and places ODT within the broader conceptual framework for studying diets (*see also* Chapter 2); (iii) summarizes existing theory and data on the effects of competitors, predators, parasites and prey behaviour on forager diets. In each section attention is drawn to areas that deserve further study.

In many cases, three trophic level effects are discussed. The levels are: predators, foragers and prey. Predators attack foragers that, in turn, attack prey. The focus is on how forager diets are influenced by other foragers, by predators and by prey behaviour.

OPTIMAL DIET THEORY AND THE COMPONENTS OF PREFERENCE

A conceptual framework is needed for organizing thoughts on how ecological interactions might influence diets. What are some known effects of ecological interactions on forager diets? Competition for food reduces food availability; i.e. it reduces forager encounter rates with prey. If competition reduces the availability of some prey more than others, this should directly alter diets. Furthermore, competition theory suggests that when competitors reduce prey availability, competitors should respond by altering their attack probabilities; for example, foragers might avoid attacking prey that are the preferred prey of a competitor (MacArthur & Levins 1967; Abrams

1983). Prey actively avoid encountering foragers, and use escape strategies to prevent being captured or consumed (Edmunds 1974); i.e. prey behaviour influences forager encounter rates with prey, and capture and consumption success. If prey differ in behavioural response to predators, this will influence diets. Foragers also respond to their own predators often by shifting into safer habitats and reducing forager activity (Sih 1987; Lima & Dill 1990). These shifts might affect diets by altering forager encounter rates with different prey. Thus encounter rates, attack probabilities given an encounter, capture and consumption success all influence preference and can all be altered by ecological interactions.

Accordingly, four components of preference can be distinguished: encounter, attack, capture and consumption (Fig. 9.1); it might also be useful to recognize additional components, e.g. detection or recognition. Preference is proportional to K/N (Chesson 1983), where K is the rate of consumption of a given prey type, and N is the number of that type in the environment. Preference is the product of the four components: (i) relative encounter rate, E/N; (ii) attack probability given an encounter, A/E; (iii) capture success given an attack, C/A; and (iv) consumption probability given a capture, K/C. E, A, and C are numbers of encounters, attacks and captures per unit time.

ODT makes predictions on only one of these four components: active predator choice (attack probabilities). If prey do not vary much in the other three components, then ODT should provide a useful framework for analyzing preferences. If, however, the other factors vary among prey, then they should also play an important role in determining diets. Note, however, that even if the other three components of preference are important, ODT still provides a potentially useful integrating role. ODT makes predictions on interactions among the different components; e.g. on how encounter, capture or consumption probabilities ought to influence attack probabilities.

The standard predictions of classical ODT are (Stephens & Krebs 1986; *see also* Chapter 1): (i) predators should attack more profitable prey; (ii) the tendency to specialize on more profitable prey should increase with an increase in encounter rates with these more profitable prey; or, more specifically, as encounter rates with more profitable prey increase, less profitable prey should be dropped from the diet; and (iii) attack probabilities on less profitable prey should not be affected by their own encounter rate.

Classical ODT defines prey profitability as the expected net energy benefit gained from attacking a prey item divided by time spent handling that item. Although often neglected by empiricists testing ODT, capture

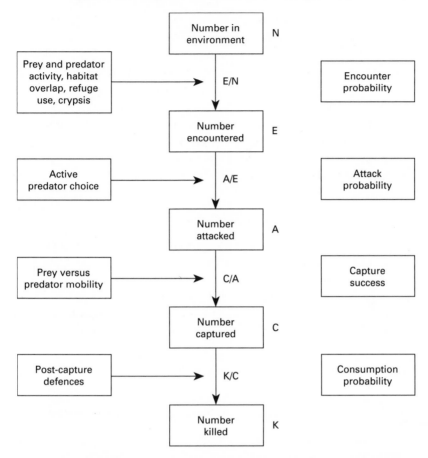

Fig. 9.1 The stages of a predator–prey interaction. On the right are four components that contribute to predation rates and preference. On the left are prey and predator attributes that often have major effects on each component. See the text for a more detailed description.

and consumption probabilities are part of the calculation of prey value: prey profitability is $(ec - (1 - c)\,x)/h$, where e is the net energy gain if prey are captured and consumed, c is the probability that an attack results in consumption (i.e. $(C/A)\,(K/C)$), x is the energy cost if prey are attacked but not consumed, and h is handling time. An increase in capture success (C/A) or consumption success (K/C) should thus not only directly increase preference (K/N) for a prey type, but should also indirectly increase preference by increasing attack probabilities (A/C).

Many of the effects of ecological interactions on preferences can be handled by the above framework: competition, predation risk or prey behaviour can influence diets by altering encounter rates and prey value, and thus attack probabilities for different prey. Some interactions, however involve costs or benefits that cannot be evaluated within the classical ODT framework. For example, some prey can only be captured by foragers if foragers put themselves under predation risk, while other prey can be handled and consumed with less risk. Classical ODT does not account for mortality considerations that clearly should influence diets. Recent theory, however, provides suggestions on how adaptive foragers ought to balance risk and energetic needs to choose a fitness-maximizing (as opposed to energy-maximizing) diet (Chapter 2).

COMPETITION AND DIETS

The main effect of competition on diets comes via effects on prey availability and thus forager encounter rates with some or all prey types. One mechanism is simple depletion of prey by competitors; i.e. I get less to eat because you ate some prey that I could have otherwise eaten. This is referred to as exploitative, or scramble, competition. Alternatively, the mechanism can involve direct conflict (e.g. aggressive interactions)

Table 9.1 Summary of the effects of competition on forager diets

---

1 *Exploitative competition*
Food limitation causes divergence in habitat use and associated divergence and narrowing of diets. Theory ($++$), non-experimental field work ($++$), field experiments ($+$), mechanistic field work ($+$).

2 *Interference competition*
a Dominant species alters the diet of subordinate species by excluding the subordinate from preferred habitats. Theory (0), field experiments ($+$).
b Dominant steals food from subordinate; subordinate thus prefers foods that are less likely to be stolen. Theory ($+$), non-experimental field study ($+$), experimental work (0).
c Aggressive interactions alter diets by reducing encounter rates for food, and by interrupting foraging bouts. Theory ($+$), laboratory experiments ($+$), field work (0).
d Aggressive interactions alter diets by reducing the value of a contested resource. Theory (0), empirical work (0).
e Superparasitism: adult 'diet' (oviposition) choices reflect avoidance of larval interference competition. Theory ($++$), experiments (0).

3 *Indirect community interactions affect diets.* Theory (0), experiments ($+$).

$++$, well studied.
$+$, a few examples.
0, essentially no studies.

among competitors; i.e. I get less to eat not because you ate the food first, but because you chase me away from the food. This is interference, or contest, competition. Effects of both exploitative and interference competition on diets are summarized in Table 9.1.

Exploitative competition

Modern theory on the effects of exploitative competition on diets began with Robert MacArthur (MacArthur & Levins 1967), who also pioneered optimal diet theory (MacArthur & Pianka 1966). His grand vision was to use optimality theory to predict behaviours that determine species interactions that explain community patterns. Early theory noted that competition could either increase or decrease diet breadth (range of food items in the diet), depending on patchiness of the food supply. Within a single patch, ODT predicts that food depletion should result in broader diets: if high-value foods are abundant, foragers should specialize on them, but if high-value foods are scarce, then foragers must broaden their diets to include lower value prey. If, however, multiple patches are considered, then competition can result in narrower diets. In response to competition-induced food depletion, foragers should restrict their habitat use to patches that are comparatively rich. If each patch contains a relatively narrow set of food types and if different patch types contain different food types, then competition-driven divergence in habitat use can result in divergence of diets into different specializations. This has been called the 'niche compression' hypothesis (MacArthur & Wilson 1967). Recent theory clarifies the conditions under which competition should either expand or contract optimal diet breadths (Abrams 1990).

How do real foragers change their diets in response to competition? If foragers do not alter their patch use in response to competition, then, as predicted by ODT, competition can result in broader diets. The likely mechanism can be reduced encounter rates with prey due to prey depletion, but might also involve a 'first-come–first-served' effect on feeding motivation. For example, in a laboratory experiment, Dill and Fraser (1984) compared the diets of juvenile coho salmon in the presence and absence of a mirror (simulating a competitor). They showed that without any prey depletion, the presence of a mirror increased the forager's attack tendency, thus resulting in a broader range of prey items in the forager's diet.

In nature, however, foragers usually have the option of shifting their patch use and diets. Many field studies suggest that foragers respond to competition by showing reduced niche overlap, narrower patch use and more specialized diets (Schoener 1986). For example, most studies comparing diets in seasons of high versus low food availability (when competition is presumably more severe) find dietary divergence and narrower diets during lean times, relative to seasons of plentiful food (Schoener 1982).

A classic example involves 'Darwin's finches' on the Galapagos islands (Grant 1986). In the early dry season, when food is abundant, all ground finches (*Geospiza* spp.) feed on soft, easy-to-handle seeds, fruits and caterpillars: coexisting finches eat similar prey, all feeding on a relatively broad range of high-value prey. Later in the dry season, when food is scarce, different finch species specialize on different diets reflecting differences in bill characteristics. Finches with small bills feed primarily on small, soft fruits; those with medium-sized bills feed on medium-sized fruits; and cactus finch feed almost exclusively on cactus nectar and pollen (Smith *et al.* 1978). Similar patterns emerged when comparing finches on different islands. When *G. difficilis* is found without *G. fuliginosa*, it feeds on pollen, nectar, and seeds on both the ground and in vegetation. The same is true for *G. fuliginosa* when it is found without *difficilis*. When, however, the two species are found on the same island, they show very little diet overlap (Schluter & Grant 1982; their Fig. 1). This phenomenon is termed competitive character displacement; when a competitor is present, a forager's diet is displaced away from that of its competitor.

The work on Galapagos finches does not involve experimental manipulations of competitors or resources. Surprisingly few field studies have experimentally manipulated competitor density to examine effects of competition on diets. Schoener (1983) reviewed over 150 experimental field studies of competition and found only four studies that addressed effects on diets. Of these four studies, only two found significant effects of competition on diets (Werner & Hall 1976; Pacala & Roughgarden 1985). Since then, a few more experimental studies have looked for diet shifts in response to competition (e.g. Spiller & Schoener 1990; Bergman 1990; Persson & Greenberg 1990); however, more studies are clearly needed.

To understand the adaptive mechanisms determining effects of competition on diets, one should use optimal foraging theory to generate quantitative predictions on the conditions where foragers ought to shift their habitat use and diets in response to competitors. Collaborative work by Earl Werner, Don Hall, Gary Mittelbach and colleagues exemplifies this approach. They studied competition and the foraging behaviour of a guild of sunfish inhabiting Michigan lakes. When held one species at a time, three species used similar habitats and fed on similar prey. When held together, they fed on different prey in different habitats (Werner & Hall 1976). Werner and colleagues used laboratory experiments to estimate the parameters of optimal foraging theory (encounter rates, handling times, capture success, energy content) for major prey types, as a function of prey and forager size, prey density, and habitat type (e.g. open water vs. vegetation vs. benthos; Werner 1977; Mittelbach 1981). These data in combination with field samples on natural prey densities and prey and predator size distributions in each habitat were used to generate predictions on optimal habitat use and diets. These predictions successfully explained seasonal habitat and diet shifts (Werner & Hall 1979; Mittelbach 1981), niche shifts in response to experimental manipulations of competitors (Werner & Hall 1977) and patterns of competitor coexistence (Werner 1977).

More recently, the same basic approach has been used to illuminate competitive dynamics and foraging behaviour of similar fish in Swedish lakes (Persson 1990; Persson & Greenberg 1990). These studies show that optimal foraging theory can provide a mechanistic understanding of how species interactions influence diets and habitat use. Despite these successes, few other research programmes have followed this approach (Pulliam 1985; Benkman 1987).

Interference competition

Interference competition involves direct conflicts among competitors (e.g. territoriality, dominance hierarchies, fights). The effects of interference on diets depend heavily on the nature of the interaction; e.g. on whether interference involves lengthy contests as compared to clear dominance with very little time devoted to aggression. Game theory suggests that the time devoted to an aggressive interaction and the outcome of the interaction depends on the existence of asymmetries (i.e. differences) between competitors (Maynard Smith 1982; Parker 1984). If there are asymmetries, then game theory predicts that these asymmetries will often be used to settle disputes with little overt aggression. Examples of asymmetries include differences in size/strength (usually, large wins, small retreats without a fight), ownership/residence (resident wins) and hunger (hungrier competitor wins). If, however, there are no clear asymmetries (e.g. similar-sized contestants with no established ownership) or conflicting asymmetries (e.g. smaller resident versus larger intruder) then the result can be lengthy, costly fights.

In asymmetric interactions with clear dominance/subordinate relations, the dominant individuals (and species) often force subordinates into sub-optimal habitats (Schoener 1983). Schoener's (1983) review found only one experimental field study that addressed effects of interference competition on diets (Pacala & Roughgarden 1985). An example of a more recent, relatively complete study on interference and diets is work by Dickman (1988, 1991), who studied three pairs of small mammalian insectivores (two pairs of species in Australia, one in England). The larger species of each pair was dominant over the smaller one; that is, the smaller species actively moved away from the presence (or odour) of the larger species. Field data on diets, microhabitat use, and prey availability, along with field experimental removals of the larger, dominant species showed that: (i) both small and large foragers prefer larger prey; (ii) although both species tended to eat larger prey than would be expected based on availability in their respective foraging microhabitats, when together, relative to small foragers, large foragers ate larger prey because large foragers used micro-habitats that contained more large prey; (iii) small foragers avoid microhab-itats with larger prey only because they are excluded by interference with large foragers. Thus interference caused habitat shifts that also forced sub-ordinates into feeding on smaller prey than they would otherwise prefer.

Dominant individuals do not always exclude subordinates from pre-ferred feeding microhabitats. In some systems, dominants instead 'para-sitize' subordinates. For example, dominants follow subordinates and displace them after subordinates have located food, or in some cases steal food after it has been captured by subordinates (e.g. Rohwer & Ewald 1981; Barnard 1984). The act of stealing food from other individuals (hosts) is termed kleptoparasitism. For example, hyenas steal from leop-ards and hunting dogs; lions steal from hyenas (Kruuk 1972). Gulls steal from plovers and other gulls (Thompson & Barnard 1984; Thompson 1986; Hockey & Steele 1990). Wasps and scorpion-flies kleptoparasitize spider webs (Vollrath 1984). Larger ants steal from smaller ants (Sav-oleinen 1991). Reviews of kleptoparasitism can be found in Brockmann and Barnard (1979), Barnard (1984) and Vollrath (1984).

The effect of the threat of kleptoparasitism on host diets can be pre-dicted by ODT. Kleptoparasitism reduces the host's probability of con-suming a given prey item given a capture (K/C), and thus reduces prey value. If all prey types are equally likely to be stolen, then kleptopara-sitism should not alter the host's diet. If, however, high-ranking prey (those with high profitability in the absence of kleptoparasitism) are more likely than others to be stolen, then hosts should broaden their diet to include low ranking prey, and if kleptoparasitism is severe enough, hosts might even switch to prefer prey that have low profitability in the absence of kleptoparasitism. For example, Savoleinen (1991) found that a subordinate wood ant preferred larger food items in the absence of a dominant wood ant, but that dominants tended to take larger food items away from subordinates. Accordingly, in the presence of the dominant ants, the subordinates primarily collected smaller food items. Thompson and Barnard (1984) explicitly applied ODT to analyse diet choice by plovers (foraging for worms) that are kleptoparasatized by gulls. In the absence of gulls, the most profitable prey for plovers are small to medium sized worms. Gulls almost exclusively steal medium-large worms from plovers; the gull's kleptoparasitic diet choice can be explained by ODT (Thompson 1986). Because gulls steal medium-sized worms, in the presence of gulls, small worms are most profitable for plovers. Observed plover diet choice showed a qualitative fit to the predictions of ODT that accounts for kleptoparasitism.

Aggressive interactions take time away from foraging. This is generally true regardless of whether the focus is on the dominant or subordinate individual. The loss in time can be accounted for as either a reduction in overall encounter rate over an entire foraging period (made up of a series of foraging bouts interrupted by other needs), or as a source of interruptions that reduce the length of each foraging bout. Classical ODT predicts that a reduction in encounter rate (particularly with high value prey) should make foragers less choosy. Shorter foraging bouts should also make foragers less choosy; i.e. if the foraging bout is likely to be interrupted soon, the forager cannot afford to wait for a higher value food item (Lucas 1985). Jaeger *et al.* (1983) experimentally addressed the prediction that interference competition among red-backed salamanders makes competitors less selective (between large and small flies). They contrasted active predator choice by dominant, territorial individuals faced with six levels of intrusion ranging from low (competitor's odour only) to medium (presence of a familiar intruder) to high (presence of an unfamiliar intruder) competition. Exploitative competition was excluded by replacing prey as they were consumed. As predicted by theory, in low-competition regimes, salamanders actively chose larger, more profitable flies; while in high-competition regimes, salamanders were non-selective.

Contests involving similar contestants or those with conflicting asymmetries often result in particularly long, costly fights (Austad 1983; Crespi 1988). According to theory, contestants should be willing to persist longer in fights for more valuable resources (Maynard Smith 1982). This increases the handling time and thus decreases the value of prey that are otherwise high in profitability and so broadens the optimal diet. By analogy with optimal patch use under competition, at the ESS, contests should make all prey equally profitable. Explicit theory on this should prove useful.

A particularly severe type of interference involves competition among insect parasitoids sharing a host. Adult female parasitoids lay eggs on hosts. Larvae typically burrow into and consume the host's tissues, almost always killing it. Superparasitism occurs when more than one parasitoid is present on one host. In many cases, competition between parasitoids produces only one survivor; i.e. competition is winner takes all, death to all losers. The mechanism involves interference; e.g. winners can release toxins that poison losers or winners bite losers, thus inducing host encapsulation of the loser (Lawrence 1988). 'Diet choice' is by the ovipositing female. Conventional wisdom suggested that because of the severity of competition (particularly because the first larva usually wins), parasitoids should always avoid superparasitism. In reality, superparasitism often occurs (Speirs *et al.* 1991). More recent theory has turned the logic upside-down. Iwasa *et al.* (1984) noted that because search time for the next host is usually much longer than handling time on a given host, ODT predicts that, if adult females have no time- or egg-number constraints, females should avoid superparasitism only if the larva would have virtually no chance of survival. Constraints on egg production (e.g. limited number of total eggs, or egg production rate) favour avoidance of superparasitism (Iwasa *et al.* 1984; Charnov & Stephens 1988), while time constraints (finite lifetime) favour superparasitism, particularly for old females (Mangel 1987; i.e. if death is imminent, eggs must be laid now). To date, experimental studies of superparasitism have not tested ODT predictions (Speirs *et al.* 1991).

Facilitation

Finally, it is worth noting that foragers that share resources are not always competitors; sometimes they increase each other's foraging rates. This is called facilitation. A plausible mechanism involves changes in prey behaviour. Prey responses to predator A can make prey more susceptible to predator B; Charnov *et al.* (1976) termed this resource enhancement. Some foragers specialize in taking advantage of resource enhancement; i.e. some foragers feed heavily on prey that have been exposed by the action of other foragers. For example, ant-birds follow ants and attack prey exposed by ant activity (Willis 1968), and cattle egrets follow ungulates for a similar reason (Burger & Gochfeld 1987).

Interactions among foragers can involve a mixture of interference and enhancement. For example, Soluk and Collins (1988) studied a system with predatory fish (sculpin), predatory insects (stoneflies) and two mayfly prey (*Baetis* that are often found on the tops and sides of stones, and *Ephemerella* that are usually found under stones). Their study contrasted interactions between sculpin and stoneflies when offered one prey type at a time. The presence of sculpin forced stoneflies to hide under rocks; i.e. sculpin interfered with stonefly foraging. This, however, increased stonefly habitat overlap with, and thus feeding rate on *Ephemerella*. While hiding under rocks, stoneflies forced some *Ephemerella* out into the open. This enhanced sculpin feeding rates on *Ephemerella*. The observed habitat shifts suggest that both predators should increase the other's preference for *Ephemerella*. Soluk and Collins (1988), however, did not directly test this prediction; i.e. they did not observe forager diets when foragers were offered a choice between the two prey types. Thus, Soluk and Collins (1988) documented the existence of complex, indirect, behaviour-mediated species interactions, but did not directly quantify the effects of these interactions on forager diets. In general, while the study of complex interactions has generated considerable excitement among community ecologists, to date, no studies have explicitly examined effects on diets.

PREDATION RISK AND DIETS

The study of effects of predation risk on various aspects of foraging behaviour has burgeoned recently. Reviews by Sih (1987) and Lima and Dill (1990) list numerous studies showing that risk influences where and when foragers feed, forager movement and activity, vigilance, forager behaviour while handling prey and social behaviour. Each of these responses to risk can indirectly alter forager diets. In addition, even if all else is unchanged, risk ought to influence forager diets because some prey items are riskier than others to handle (e.g. some prey require more intensive handling that interferes with vigilance). Below, some of the main effects of predation risk on forager diets are discussed and summarized in Table 9.2).

Predation risk often alters forager habitat use; i.e. risk often causes foragers to increase their tendency to search for food in safer habitats

Table 9.2 Summary of the effects of predation risk, parasitism and prey behaviour on forager diets

---

1   *Predation risk*
   a Risk alters forager habitat use and thus associated diets. Theory (++), field and laboratory experiments (+).
   b Risk reduces forager activity and thus broadens diets. Theory (+), laboratory experiments (+), field work (0).
   c If handling is riskier than search, the foragers should be more choosy, and vice versa. Theory (+), laboratory experiments (+), field work (0).
   d With predators present, foragers should prefer safer prey. Theory (+), laboratory experiments (+), field experiments (0).
   e Risk increases the forager's tendency to carry large prey to safety for handling. Theory (+), field experiments (+).
   f Risk for nestlings induces parents to be less choosy about the value of food brought back to the nest. Theory (+), field experiments (+).

2   *Uncertainty about predation risk influences forager sampling behaviour and thus diets.* Theory (0), experiments (0)

3   *Parasitism*
   a Increases forager nutritional demands and thus influences diets. Theory (+), experiments (+).
   b Foragers avoid foods that harbour parasites, prefer foods that have anti-parasitic properties. Theory (0), non-experimental work (+), experiments (0).

4   *Prey behaviour*
   a Prey avoid encountering foragers, this influences forager diets. Theory (0), field and laboratory studies, including experiments (++).
   b Prey attempt to escape when attacked, this influences forager diets. Theory (+), field and laboratory studies, including experiments (++).
   c We do not, however have an integrative framework for addressing the interacting effects of adaptive prey and forager behaviour on the different components of the predator–prey interaction that together determine diets.

++, well studied.
+, a few examples.
0, essentially no studies.

(Sih 1987; Lima & Dill 1990). If different prey types are found in different habitats, then the habitat shift will indirectly cause a change in diets. For example, Werner *et al.* (1983) showed that the presence of predatory largemouth bass caused small bluegills to shift from a diet of zooplankton in the dangerous, open-water zone to macroinvertebrates in the safer, nearshore vegetation. Other examples of predator-induced habitat shifts causing diet shifts include studies on sea urchins (Vance & Schmitt 1979) and moose (Edwards 1983).

Risk also causes changes in foraging behaviour within a habitat; e.g. risk often causes foragers to reduce foraging activity (Sih 1987; Lima & Dill

1990). This should reduce encounter rates with predators and make foragers less conspicuous to predators. A cost of reduced activity, however is a decrease in the forager's encounter rates with its own prey. According to ODT, risk should thus cause foragers to broaden their diets. For example, Sih and Moore (1990) studied the effects of predatory green sunfish (or actually the smell of sunfish) on active predator choice between amphipod and isopod prey, by salamander larvae. In the absence of sunfish, salamanders fed almost exclusively on amphipods. The addition of sunfish smell reduced salamander and prey activity, thus decreasing salamander encounter rates with both prey types. Apparently in response to decreased encounter rates, salamanders increased their attack probability (A/E) on both prey types (particularly on isopods) and fed non-selectively.

Alternatively, diet choice within habitats can be determined by variations in risk associated with different prey. Pearson (1976) modified classical ODT to include mortality risks associated with search and handling, and differences among prey types in risk while handling. His work, however, was never published. Recently, Lima (1988a), Gilliam (1990) and Godin (1990) published models on the effects of risk on diets, each adding significant insights to Pearson's work. Unlike classical ODT which assumes a goal of energy maximization, the risk-diet models assume that the forager's goal is to maximize survival by balancing risks of predation against risks of starvation.

In the simplest view, foragers ought to minimize predation risk while meeting a minimal energy requirement. Using this assumption, Gilliam and Fraser (1987) showed that the optimal strategy is one that maximizes the ratio of feeding rate/mortality rate. Prey rankings should then be based not on (net energy gain/handling time), but on (net energy gain/total risk while handling). Total risk while handling is risk per unit handling time multiplied by handling time (Pearson 1976; Gilliam 1990; Lima (1988a) came to a similar criterion using a more complex model); predators should, therefore, prefer low profit, safe prey over high profit, dangerous prey if the safer prey have a higher benefit/risk ratio.

Gilliam (1990) examined the effects of relative predation risk associated with search versus handling. If handling is more dangerous than search (e.g. if handling interferes with vigilance, but search does not), then risk should make foragers more choosy; i.e. if handling is relatively dangerous, then foragers should reject some prey that they would otherwise take, and instead continue searching for other higher value prey. On the other hand, if search is more dangerous than handling, then risk should make foragers less choosy.

Pearson (1976), Lima (1988a) and Godin (1990) explicitly considered the effects on optimal diets of variation among prey in associated risk. The interesting case is where energetic profitabilty conflicts with safety. They predicted that in a two-prey system, as risk increases, foragers should switch from a preference for the energetically more profitable prey, to non-selective foraging, to a preference for the safer prey. The rationale is that if risk is low, there is little need to worry about it; thus prey with high energetic profitability are preferred. In contrast, if risk is high, diet choice is dominated by risk considerations; thus safer prey are preferred.

The models also examined effects of hunger on dietary responses to risk. Godin (1990) used a dynamic programming model that explicitly accounts for energy reserves as a state variable to show that when a forager's energy reserves are high (e.g. the forager is satiated), it should 'play it safe' and specialize on safer prey, whereas when energy reserves are low, the forager should be less choosy. Gilliam (1990) reached similar conclusions by varying threshold energetic needs in a static optimization model. In Chapter 2, Houston gives a full description of dynamic versus static optimization models.

Finally, the models note that predation risk can explain partial preferences. Classical ODT predicts that foragers should have attack probabilities of either zero or one; i.e. they should not show partial preferences. In fact, foragers almost always show partial preferences. Gilliam's model (1990) generated partial preferences in the following way. Imagine that a diet consisting of the two highest ranked prey (ranked by benefit/risk) minimizes risk, but is insufficient to satisfy energy needs. To satisfy its energy needs, the forager must broaden its diet to include prey type 3. The forager, however, should not eat all type 3 prey that are encountered. It should eat only as many as necessary to satisfy its energetic needs. Pearson (1976) and Lima (1988a) derived similar, though somewhat more complicated, results using more complex models.

Only a few empirical studies include data that are relevant for testing the predictions of the above models. Some studies focus on sit-and-wait foragers, while others examine foragers that actively search for prey. Sit-and-wait foragers search for prey from the safety of a refuge, but must temporarily leave the refuge to pursue and capture prey; i.e. handling is riskier than search. For example, Dill and Fraser (1984) studied juvenile salmon that hover at a stationary spot (presumably, a relatively safe site) and dash out to intercept flies as they drift by in a steady current. Gilliam's model (1990) predicts that in this scenario, risk should tend to

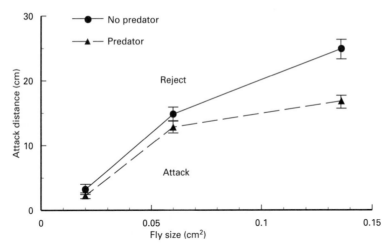

Fig. 9.2 Effects of predation risk on the attack distances of juvenile salmon on three size classes of prey. Prey value is influenced by prey size and distance from the salmon's safe site. If the distance from the safe site to prey is above a line, prey are rejected; below a line, prey are attacked. The region between the lines is one in which salmon attack prey in the absence of predation risk, but reject prey in the presence of predation risk; i.e. salmon are choosier in the presence of risk. Data are means ± SE. Modified from Dill and Fraser (1984).

make foragers more choosy. Here, prey value depends on both energetic profitability and risk, which is proportional to distance from the safe spot. Figure 9.2 shows that when the salmon were exposed to a model predator, they reduced their attack distance, particularly for larger, more energetically profitable prey: that is, with a predator present, the forager rejected large, distant prey that were judged to be acceptable in the absence of predators. Interestingly, the salmon's size selectivity actually decreased in the presence of predators. Without a predator, salmon preferred large flies; whereas, with a predator present, salmon preferred prey that were close to cover, but were less selective relative to prey size. Put another way, in the absence of predation risk, the forager's diet choice was based primarily on energetic profitability, while in the presence of risk, their choices were based more on safety considerations. Similar patterns were observed for another species of salmon (Metcalfe *et al.* 1987a).

Foragers that actively search for food outside their refuge typically face risks during both the search and handling phases. Different prey, however, might differ in their associated risk. Theory predicts that in the absence of predators, actively searching foragers should prefer energetically profitable prey; however, when predation risk is high, these foragers

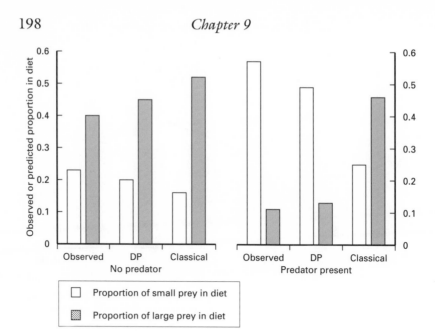

Fig. 9.3 A comparison of predicted and observed diets of guppies selecting among three classes of zooplankton, *Daphnia*. Shown are the proportions of small and large prey in the diet; medium-sized prey always made up about 30% of the diet. DP = diet predicted by a dynamic programming model that accounts for predation risk. Classical = diet predicted by the classical optimal diet model that does not account for risk. Modified from Godin (1990).

should prefer safer prey. A key determinant of risk associated with handling prey is the effect of handling on vigilance. In many systems, handling interferes with vigilance so that smaller prey that require shorter handling times are safer. Godin's (1990) study of guppy diet choice among three size-classes of zooplankton provides an unusually detailed test of the effects of risk on the diet of an active forager. He calculated net energetic values and handling times and observed all of the relevant components of preference (encounter rates, attack probabilities and capture success). He then compared the diet predictions of classical ODT (that does not account for mortality risk), with a dynamic programming (DP) model that accounts for both predation risk and the risk of starvation (Fig. 9.3). His experiment, however, used a predator behind a glass partition; thus he could only guess at the guppies' perceived risk of predation. In the absence of predators, both models adequately predicted diet choice. Only the DP model, however, accurately predicted the effects of predators. The presence of a predator caused guppies to decrease their encounter rates and capture success on medium- and large-sized prey. Based on energetics alone, the ODT model predicted that guppies should have showed no active choice, and thus should have eaten more large prey because large prey were encountered more frequently. The DP model, however, predicted that guppies under risk should actively choose smaller, safer prey. This is, in fact, what guppies did. With a predator present, guppies rarely ate large prey, even though they encountered them most frequently, and primarily ate smaller, safer prey, even though they encountered them relatively rarely.

The costs and benefits of active foraging are influenced by the forager's social milieu. Individuals in groups enjoy 'safety in numbers' via mechanisms such as increased group vigilance, dilution of risk, and predator confusion (Pulliam and Caraco 1984). Individuals in larger groups also face stronger resource competition. Given these trends, Lima (1988b) predicted that as group size decreases, the diets of dark-eyed juncos ought to reflect safety considerations more and energetic needs less. In his system, smaller food items were energetically more profitable, but more dangerous because they had to be eaten with the forager's head down, while larger items were energetically less profitable, but safer because they could be eaten with the head up. His prediction was thus that birds in large groups ought to prefer smaller, profitable foods, while birds in small groups ought to prefer larger, safe foods. Observations on juncos corroborated these predictions.

Some foragers can mix active search with handling in safety, that is, foragers can carry food back to be handled in refuges. The trade-off is between the safety of handling food in refuge versus the energetic and time costs of carrying food rather than simply eating it where it is found. Theory predicts that foragers should carry a higher proportion of items to the refuge, if the food items are larger and closer to safety. Experiments with grey squirrels and a dozen bird species showed that the prediction on food size and carrying is usually corroborated, but the prediction on distance to cover and carrying often fails (Lima *et al.* 1985; Lima & Valone 1986; Valone & Lima 1987). Lima and Valone (1986) also predicted that the option to carry foods also alters the optimal diet. They studied a situation where grey squirrels chose between smaller, more profitable versus larger, less profitable food items. As predicted by theory, in situations where carrying is favoured, squirrels carried food more often; given that they only carried large items in these situations, squirrels increased their preference for large items.

The work on carrying food to refuge illuminated another aspect of foraging behaviour that can respond to predation risk: handling time *per se*. The trade-off is that for a given prey item, longer handling times are risky, while shorter handling times might lead to inefficient digestion. The usual trend is that when foragers handle prey outside the refuge, they handle them very quickly (and thus probably suffer poor digestive efficiency), but when the same foragers handle the same prey in safety, they handle them slowly. If prey vary in how easily they can be handled quickly and yet still efficiently, this ought to play a role in determining optimal diets under risk. This possibility has not yet been tested.

A variation on carrying prey involves central place foragers; e.g. animals that bring food to young at a nest. Two studies have looked at how parents feeding and protecting young at a central place alter their diets in response to predation risk. If the main effect of risk is to induce parents to return to the nest more frequently (i.e. after briefer foraging bouts), then theory suggests that risk should make parents less choosy (Lucas 1985). If we account explicitly for the fact that searching for food by the parent is more dangerous to nestlings than is food handling (feeding of young), then Gilliam (1990) predicts both that parents should search a smaller proportion of the time, and that parents should be less choosy. Freed (1981) compared the diet choice of parental house wrens *before* vs. *after* exposure to a predator on nestlings (a tethered snake). The snake (if released from its tether) could kill nestlings, but also could be driven away by parents. As predicted by theory, after exposure to the snake, parents took shorter foraging bouts, foraged closer to the nest, and returned with smaller prey. Martindale (1982) observed similar patterns in Gila woodpeckers.

In sum, existing theory predicts that predation risk ought to alter both diet breadth and diet preferences. The effects of risk on optimal diets depends on the specific mechanism; e.g. on whether handling or search is riskier and on whether risk conflicts with energetic profitability. To date, few empirical studies have tested these predictions. More tests should provide further insight.

PREDATION RISK, UNCERTAINTY, ERRORS AND DIETS

Environments vary in space and time. Organisms usually experience uncertainty about environmental variation. Recent interest has focused on how foragers assess changes in the distribution, abundance and values of foods (Stephens & Krebs 1986; Shettleworth, Chapter 4; Provenza, Chapter 5). For foragers, one consequence of uncertainty is that they must absorb some short-term reductions in feeding rate while sampling their environment. They do not know that a patch or food item is low in value until they try it. Foragers also experience uncertainty about predation risk. The cost of errors while sampling for predators, however, can be much greater than the cost of errors while sampling for food. To assess risk, foragers might often have to expose themselves to dangerous predators. Under a broad set of conditions, the optimal behaviour is to not sample, but instead to stay in refuge (Sih 1992). This constraint should also restrict the ability of foragers to monitor changes in food availability. Thus predation risk can produce suboptimal foraging by reducing forager responses to a changing food environment. Although a few studies have demonstrated that foragers often 'over-respond' to predators (i.e. stay in refuge even after foragers have been removed; Lima and Dill 1990; Sih 1992), no studies have looked at the consequences for long-term shifts in diet or patch choice.

In a similar vein, predation risk can simply cause foragers to make mistakes in diet choice. Metcalfe *et al.* (1987b) observed the diets of juvenile salmon given a choice between edible and inedible foods. They found that the frequency of attacks on inedible foods (i.e. recognition errors) increased after the salmon were exposed to a model predator.

PARASITISM AND DIETS

Lozano (1991) recently reviewed several ways in which parasitism or the threat of parasitism might influence host diets (Table 9.2). One mechanism involves indirect effects via the influence of parasite-induced stress on host behaviour. Because parasites take energy from hosts, the hosts' nutritional requirements are increased. To compensate for this, hosts have been shown to be less responsive to predators (Giles 1987), and to shift their diets (Milinski 1984). Other mechanisms are direct responses to parasitism: (i) potential hosts can avoid foods that are sources of parasites (this is the common wisdom behind human avoidance of under-cooked pork): (ii) hosts can prefer foods that harm parasites; e.g. natural medicines or foods that alter the host's internal environment so that it is less suitable for parasites. For example, in areas with schistosomes, baboons eat leaves and berries of a shrub that is toxic to schistosomes; in areas that are free from schistosomes, baboons do not consume this shrub (Philips-Conry 1986). To date, no studies have provided detailed data on trade-offs that might determine host dietary responses to parasitism. For example, is it often the case that foods that are high in profitability are also more likely to harbour parasites? Do foragers avoid profitable, but parasitized prey? Is the avoidance of parasites flexible in degree and is this mediated by the costs and benefits of such avoidance? Along similar lines, are anti-parasitic foods often low in energetic profitability? If so, is host preference for anti-parasitic foods mediated by associated costs and benefits?

PREY BEHAVIOUR AND FORAGER DIETS

As emphasized earlier, a forager's diet is not necessarily controlled primarily by the forager, but can be heavily influenced by prey characteristics (Table 9.2). This is true for all stages of the predator–prey act. In the encounter phase, predators search for prey and prey hide from predators. If some prey hide more effectively than others, then this variation in prey availability can be the primary factor determining forager diets. In the attack phase, predators actively choose to attack some and ignore other prey. Prey behaviour, however, can strongly influence attack decisions. For example, some foragers attack anything (within some size range) that moves. Variation among prey in tendency to move should then be the main factor determining attack probabilities. In the capture and consumption phases, predators try to subdue and consume prey, while prey try to escape (by flight or fight) from predators. The goal of this section is to: (i) identify generalities on the relative importance of prey and predator behaviour in each stage; and (ii) discuss empirical techniques for quantifying the relative importance of the different stages and behaviours.

One generality is that if prey have potentially effective anti-predator behaviours, then variation among prey in anti-predator behaviour is a major determinant of forager diets. For example, on the African plains of the Serengeti National Park, cheetah prefer male over female gazelle. Why? Fitzgibbon (1990) observed 102 hunts and 25 kills and suggested that males are killed more often because, relative to females, males are more likely to be solitary and, if in a group, males are generally less vigilant and more likely to be located at the edge of the group. All of these prey behaviours tend to make males more likely to be taken by cheetah. This example is one of several suggesting that male ungulates are generally preferred over females because males show less effective anti-predator behaviour (see Clutton-Brock *et al.* 1992 for a review). Other studies in which a terrestrial forager's diet appears to be influenced heavily by prey traits include work on eagles attacking monkeys (Struhsaker & Leakey 1990), mustelids attacking voles (Erlinge 1981; Jedrzejewska & Jedrzejewska 1990), and wasps attacking leafhoppers (Settle & Wilson 1990).

Most of the studies in terrestrial systems, however, are largely anecdotal; i.e. they do not include detailed data on prey behaviour nor on the importance of each of the components of preference.

More detailed information on the components of preference can be found in studies of freshwater predator–prey systems. For example, Osenberg and Mittelbach (1989) studied diet choice by pumpkinseed sunfish, a fish that specializes on eating snails. Using a blend of field and laboratory surveys and experiments, they estimated prey encounter rates, attack probabilities, consumption success and profitability. They used this information to predict the fish's optimal diet, and to examine the relative effects of each component on overall dietary preference. As predicted by ODT, the sunfish preferred more profitable prey, and increased their selectivity for more profitable prey when the overall environment was more productive. This suggests that active predator choice plays an important role in explaining the forager's diet. However, when Osenberg and Mittelbach (1989) compared statistically the importance of each stage in explaining variation in forager diets, they found that a model that accounts only for relative encounter rates with different prey (which depend heavily on prey size) explained the majority of the variation in forager size preferences. Accounting for variations in capture success (i.e. some prey are too large or thick-shelled to be crushed by these sunfish) significantly increased the model's ability to predict sunfish diets. Accounting for active forager choice added very little to the overall model's capacity to explain diets. Thus prey traits (the effects of snail size and species on encounter rates, and snail-crushing resistance on consumption success) were the major factors explaining why some snails are eaten and others are not. Wright and O'Brien (1984), Bence and Murdoch (1986) and Hart and Hamrin (1990) provide other detailed studies also showing that fish dietary preferences are influenced heavily by prey traits.

The above studies show that in some systems, prey traits can be important in determining forager diets. But is this generally true, or at least generally true for a particular type of forager? To address this question, Sih and Moore (1990) reviewed studies on the diets of 33 aquatic invertebrate foragers (feeding on zooplankton, aquatic insects and amphibian larvae). Studies of nine more foragers can be added to the survey (Fuller & DeStaffan 1988; Folt & Byron 1989; Peckarsky & Penton 1989; Blois-Heulin *et al.* 1990; McPeek 1990; Ramcharan & Sprules 1991). In 37 out of 42 (88%) cases, prey activity and movement were judged to play an important role in determining diets.

Capture/escape success was important in 20 (48%) diets. Active predator choice (variation in attack probabilities) was observed in only 8 (19%) of the cases. That is, foragers generally 'preferred' prey that were more active and thus encountered more often, and prey that were easier to catch. In most cases, foragers did not show active choice. Even when they showed active choice, it could be attributed to variation in prey behaviour. Peckarsky and Penton (1989) suggested that stoneflies attack anything that moves; prey that move more often in response to stoneflies are thus 'actively chosen'.

If many foragers eat whatever they run into more often and can easily catch, then ODT, which addresses active predator choice, should often fail to predict diets. This runs against an impression that ODT has been quite useful in predicting diet preferences and shifts in preferences with changes in prey abundances (Stephen & Krebs 1986). The resolution of this conflict lies in the fact that most studies that test ODT examine immobile or nearly immobile prey (e.g. seeds, flowers, dead mealworms, bivalves) that have little or no capacity to respond behaviourally to predators. Sih and Moore (1990) used a partial correlation analysis to show that prey mobility explained a significant amount of variation in the ability of ODT to explain observed diets (data were taken from a survey of 60 optimal diet studies by Stephens and Krebs (1986); see Sih and Moore (1990) for details). When prey were immobile, ODT was quite useful; however, when prey were relatively mobile, forager diets often did not fit ODT. Only 8 in 60 studies examined foragers feeding on relatively mobile prey (e.g. weasels chasing voles, or aquatic insect predators attacking aquatic insect prey).

The overall suggestion is that if prey can respond effectively to foragers (certainly a common phenomenon in nature), then prey behaviour must be integrated into theory and empirical studies on determinants of foraging behaviour. Unlike the sections on effects of competition and predation risk on diets, we do not currently have explicit theory predicting how adaptive prey behaviour ought to influence forager diets. Sih and Moore (1990) noted some situations where adaptive prey behaviour could generate patterns that resemble those predicted by ODT (but without active forager choice), and other situations where adaptive prey behaviour should result in patterns that go against ODT. Further theoretical work in this area should prove rewarding.

Finally, it is worth emphasizing that it is not the case that if prey have anti-predator behaviours, prey behaviour necessarily overwhelms active forager choice. Both might be important in shaping forager diets.

Although many studies have gathered some data on both prey and forager behaviour, no standard techniques exist for quantitatively assessing the relative importance of different behaviours. Sih and Moore (1990) suggested the use of path analysis, a multivariate statistical technique related to partial correlation analysis, to partition the relative effects of various mechanisms in determining diets. They applied the technique to a study of salamander foraging on macroinvertebrates (isopods and amphipods). Salamanders preferred amphipods. They encountered more amphipods, had a higher attack probability on amphipods and, in some conditions, had higher capture success on amphipods. Thus an 'eyeballing' of the data suggests that all components contribute to the observed preference. The path analysis (Fig. 9.4), however, showed that the key 'pathway' explaining variation in preference is the effect of encounter rates with amphipods on attack probabilities on isopods. If salamanders encountered large numbers of amphipods, then they were unlikely to attack isopods and thus

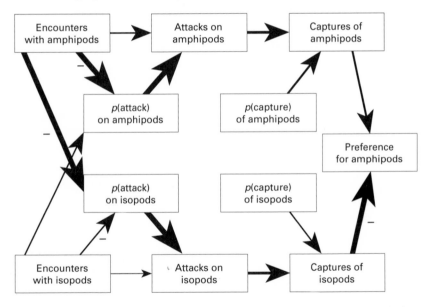

Fig. 9.4 Path diagram of factors that influence variation in preference of small-mouthed salamander larvae for amphipods over isopods. The thickness of an arrow reflects the importance of a pathway as assessed by its path coefficient. A relationship is positive unless it is indicated as negative. The main factor affecting variation in preference is the number of captures on the less preferred prey, isopods. Capture rate on isopods is influenced primarily by the attack rate on isopods. The attack rate on isopods is determined by the probability that an isopod will be attacked given an encounter which, in turn, is heavily influenced by the salamander's encounter rate with amphipods. This mechanism corroborates the predictions of optimal diet theory. Modified from Sih and Moore (1990).

preferred amphipods. This is exactly the mechanism predicted by ODT. Further use of path analysis, or related multivariate techniques should help us to better pin-point key determinants of diets, that should then be tested with further experimental work.

CONCLUSIONS

Tables 9.1 and 9.2 (*see* pages 186 and 194) summarize some main effects of ecological interactions on forager diets. Few of these effects, however, are substantiated by more than a handful of studies. The following research areas deserve further attention.

1 Although we have a considerable mass of non-experimental data suggesting that competition causes diet shifts, surprisingly few experimental studies have examined the effects of either exploitative or interference competition on diets in the field. In particular, few studies have included mechanistic studies on competition and diets. For exploitative competition, only a few studies have tested the ability of optimal foraging theory to predict diet shifts in response to competition for food. In the few cases where the approach has been applied, it has yielded powerful insights. Recent advances in ODT still need to be integrated into this approach. For interference competition, we await a more comprehensive theory on how interference might affect diets. A trail worth exploring involves the use of game theory on the dynamics of interference to predict effects on diets.

2 One interesting type of interference competition is superparasitism by insect parasitoids. Patterns of parasitism and superparasitism potentially have important practical implications relative to biological control of insect pests. A number of models make predictions on patterns of adaptive superparasitism. These models need to be tested.

3 A central theme in community ecology is the study of complex, indirect interactions among species. How does the presence of species A influence interactions between species B and C? The effects of indirect interactions on diets is an unexplored area that deserves some attention.

4 Concerning effects of predation risk on diets, we have a few models and few tests; however, many more tests are needed to illuminate general patterns to guide further theory. In particular, we need more experiments on effects of risk on diets under natural, field conditions.

5 In general, we need more work on the effects of uncertainty about predation risk on forager or prey behaviour. To date, no studies have addressed effects of uncertainty about risk on forager diets.

6 Although it seems plausible that diets might be influenced by the need to avoid/reduce parasitism, we do not have a quantitative framework for addressing these effects. We need explicit cost/benefit theory to predict patterns of parasitism-avoidance and, of course, tests of theory.

7 Although it is clear that prey behaviour often has important effects on forager diets, much work remains to be done to quantify and ultimately, predict these effects. Most studies on diets still focus on either the forager's behaviour or on the prey's behaviour; few studies quantify both. Studies that quantify both prey and forager behaviour need to adopt a statistical framework for assessing the relative effects of different components (and interactions among components) on variation in forager diets. Finally, to advance this area to a predictive science, we need cost/benefit theory on the interacting effects of adaptive prey and forager behaviour on forager diets.

In closing, it must be emphasized again that many, if not most, studies of adaptive diets do not distinguish adequately between active forager choice (as predicted by ODT) and non-random diets (preference). Diets depend on several components of the predator–prey act; ODT addresses only one of these components. Ecological interactions can influence any of the components. We need cost/benefit theory and data on each of the components (many of which will require a theory that accounts for both forager and prey viewpoints) and on how they interact. An integrative view should help us to develop a fuller understanding of both diets and how ecological interactions alter diets.

REFERENCES

Abrams P.A. (1983) The theory of limiting similarity. *Ann. Rev. Ecol. Syst.* **14**, 359–76.

Abrams P.A. (1990) Adaptive responses of generalist herbivores to competition: convergence or divergence? *Evol. Ecol.* **4**, 103–14.

Austad S.N. (1983) A game theoretical interpretation of male combat in the bowl and doily spider (*Frontinella pyramitela*). *Anim. Behav.* **31**, 59–73.

Barnard C.J. (1984) The evolution of food-scrounging strategies within and between species. In *Producers and Scroungers. Strategies of Exploitation and Parasitism* (ed. by C.J. Barnard), pp. 95–126. Chapman & Hall, New York.

Bence J.R. & Murdoch W.W. (1986) Prey size selection by the mosquitofish: relation to optimal diet theory. *Ecology* **67**, 324–36.

Benkman C.W. (1987) Food profitability and the foraging ecology of crossbills. *Ecol. Monogr.* **57**, 251–67.

Bergman E. (1990) Effects of roach *Rutilus rutilus* on two percids, *Perca fluviatilis* and *Gymnocephalus cernua*: the importance of species interactions for diet shifts. *Oikos* **57**, 241–9.

Blois-Heulin C., Crowley P.H., Arrington M. & Johnson D.M. (1990) Direct and indirect effects of predators on the dominant invertebrates of two freshwater littoral communities. *Oecologia* **84**, 295–306.

Brockmann H.J. & Barnard C.J. (1979) Kleptoparasitism in birds. *Anim. Behav.* **27**, 487–514.

Burger J. & Gochfeld M. (1987) Host selection as an adaptation to host-dependent foraging success in the Cattle Egret (*Bubulcus ibis*). *Behaviour* **79**, 212–29.

Charnov E.L., Orians G.H. & Hyatt K. (1976) Ecological implications of resource depression. *Am. Nat.* **110**, 247–59.

Charnov E.L. & Stephens D.W. (1988) On the evolution of host selection in solitary parasitoids. *Am. Nat.* **132**, 707–22.

Chesson J. (1983) The estimation and analysis of preference and its relationship to foraging models. *Ecology* **64**, 1297–1304.

Clutton-Brock T.H., Guinness F.E. & Albon S.D. (1982) *Red Deer. Behaviour and Ecology of Two Sexes.* Edinburgh University Press, Edinburgh.

Crespi B.J. (1988) Risks and benefits of lethal male fighting in the colonial, polygynous thrips, *Hophlothripis karyni* (Insecta: Thysanoptera). *Behav. Ecol. Sociobiol.* **22**, 293–301.

Dickman C.R. (1988) Body size, prey size and community structure in insectivorous mammals. *Ecology*, **69**, 569–80.

Dickman C.R. (1991) Mechanisms of competition among insectivorous mammals. *Oecologia* **85**, 464–71.

Dill L.M. & Fraser A.H.G. (1984) Risk of predation and the feeding behavior of juvenile coho salmon (*Oncorhynchus kisutch*). *Behav. Ecol. Sociobiol.* **16**, 65–71.

Edmunds M. (1974) *Defense in Animals.* Longman Inc., New York.

Edwards J. (1983) Diet shifts in moose due to predator avoidance. *Oecologia* **60**, 185–9.

Erlinge S. (1981) Food preference, optimal diet and reproductive output in stoats *Mustela erminea* in Sweden. *Oikos* **36**, 303–15.

Fitzgibbon C.D. (1990) Why do hunting cheetahs prefer male gazelles? *Anim. Behav.* **40**, 837–45.

Folt C.L. & Byron E.R. (1989) A comparison of the effects of prey and non-prey neighbors on foraging rates of *Epischura nevadensis* (Copepoda: Calanoida). *Fresh. Biol.* **21**, 283–93.

Freed L.A. (1981) *Breeding biology of house wrens: new views of avian life history phenomena.* Ph.D. thesis, University of Iowa.

Fuller R.L. & DeStaffan P.A. (1988) A laboratory study of the vulnerability of prey to predation by three aquatic insects. *Can. J. Zool.* **66**, 875–8.

Giles N. (1987) Predation risk and reduced foraging activity in fish: experiments with parasitized and non-parasitized three-spined sticklebacks, *Gasterosteus aculeatus* L. *J. Fish Biol.* **31**, 37–44.

Gilliam J.F. (1990) Hunting by the hunted: optimal prey selection by foragers under predation hazard. In *Behavioural Mechanisms of Food Selection* (ed. by R.N. Hughes), *NATO ASI series, vol. G 20*, pp. 797–819. Springer Verlag, Berlin.

Gilliam J.F. & Fraser D.F. (1987) Habitat selection when foraging under predation hazard: a model and a test with stream-dwelling minnows. *Ecology* **68**, 1856–62.

Godin J-G.J. (1990) Diet selection under the risk of predation. In *Behavioural Mechanisms of Food Selection* (ed. by R.N. Hughes), *NATO ASI series, vol. G 20*, pp. 739–69. Springer Verlag, Berlin.

Grant P.R. (1986) *Ecology and Evolution of Darwin's Finches.* Princeton University Press, Princeton, N.J.

Hart P.J.B. & Hamrin S.F. (1990) The role of behaviour and morphology in the selection of prey by pike. In *Behavioural Mechanisms of Food Selection* (ed. by R.N. Hughes), *NATO ASI series, vol. G 20*, pp. 235–54. Springer Verlag, Berlin.

Hockey P.A.R. & Steele W.K. (1990) Intraspecific kleptoparasitism and foraging efficiency as constraints on food selection by kelp gulls *Larus domini canus*. In *Behavioural Mechanisms of Food Selection* (ed. by R.N. Hughes), *NATO ASI series, vol. G 20* pp. 679–706. Springer Verlag, Berlin.

Hughes R.N. (1990) *Behavioural Mechanisms of Food Selection. NATO ASI series G, vol. 20,* Springer Verlag, Berlin.

Iwasa U., Suzuki Y & Matsuda H. (1984) Theory of oviposition strategy of parasitoids. I. Effect of mortality and limited egg number. *Theor. Pop. Biol.* **26**, 205–27.

Jaeger R.G., Nishikawa K.C.B. & Barnard D.E. (1983) Foraging tactics of a territorial salamander: costs of territorial defence. *Anim. Behav.* **31**, 191–8.

Jedrzejewska B. & Jedrzejewska W. (1990) Antipredator behaviour of bank voles and prey choice of weasels – enclosure experiments. *Ann. Zool. Fenn.* **27**, 321–8.

Kruuk H. (1972) *The Spotted Hyena: A Study of Predation and Social Behavior.* University of Chicago Press, Chicago.

Lawrence P.O. (1988) Intraspecific competition among first instars of the parasitic wasp *Biosteres longicaudatus. Oecologia* **74**, 607–11.

Lima S.L. (1988a) Vigilance and diet selection: the classical diet model reconsidered. *J. Theor. Biol.* **132**, 127–43.

Lima S.L. (1988b) Vigilance and diet selection: a simple example in the dark-eyed junco. *Can. J. Zool.* **66**, 593–6.

Lima S.L. & Dill L.M. (1990) Behavioral decisions made under the risk of predation: a review and prospectus. *Can. J. Zool.* **68**, 619–40.

Lima S.L. & Valone T.J. (1986) Influence of predation risk on diet selection: a simple example in the grey squirrel. *Anim. Behav.* **34**, 536–44.

Lima S.L., Valone T.J. & Caraco T. (1985) Foraging efficiency – predation risk tradeoff in the grey squirrel. *Anim. Behav.* **33**, 155–65.

Lozano G.A. (1991) Optimal foraging theory: a possible role for parasites. *Oikos* **60**, 391–5.

Lucas J.R. (1985) Time constraints and diet choice: different predictions from different constraints. *Am. Nat.* **126**, 680–705.

MacArthur R.H. & Levins R. (1967) The limiting similarity, convergence and divergence of coexisting species. *Am. Nat.* **101**, 377–85.

MacArthur R.H. & Pianka E.R. (1966) On optimal use of patchy environment. *Am. Nat.* **100**, 603–9.

MacArthur R.H. & Wilson E.O. (1967) *The Theory of Island Biogeography.* Princeton University Press, Princeton, N.J.

Mangel M. (1987) Oviposition site selection and clutch size in insects. *J. Math. Biol.* **25**, 1–22.

Martindale S. (1982) Nest defence and central place foraging: a model and experiment. *Behav. Ecol. Sociobiol.* **10**, 85–9.

Maynard Smith J. (1982) *Evolution and the Theory of Games.* Cambridge University Press, Cambridge.

McPeek M.A. (1990) Behavioral differences between *Enallagma* species (Odonata) influencing differential vulnerability to predators. *Ecology* **71**, 1714–26.

Metcalfe N.B., Huntingford F.A. & Thorpe J.E. (1987a) The influence of predation risk on the feeding motivation and foraging strategy of juvenile Atlantic salmon. *Anim. Behav.* **35**, 901–11.

Metcalfe N.B., Huntingford F.A. & Thorpe J.E. (1987b) Predation risk impairs diet selection in juvenile salmon. *Anim. Behav.* **35**, 931–3.

Milinski M. (1984) Parasites determine a predator's optimal feeding strategy. *Behav. Ecol. Sociobiol.* **15**, 35–7.

Mittelbach G.G. (1981) Foraging efficiency and body size: a study of optimal diet and habitat use by bluegills. *Ecology* **62**, 1370–86.

Osenberg C.W. & Mittelbach G.G. (1989) Effects of body size on the predator–prey interaction between pumpkinseed sunfish and gastropods. *Ecol. Monogr.* **59**, 405–32.

Pacala S.W. & Roughgarden J. (1985) Population experiments with the *Anolis* lizards of St. Maarten and St. Eustatius. *Ecology* **66**, 129–41.

Parker G.A. (1984) Evolutionarily stable strategies. In *Behavioural Ecology: An Evolutionary Approach* (ed. by J.R. Krebs & N.B. Davies), pp. 30–61. Sinauer Assoc., Sunderland, MA.

Pearson N.E. (1976) *Optimal foraging: some theoretical considerations of different feeding strategies.* Ph.D. dissertation, Univ. of Washington.

Peckarsky B.L. & Penton M.A. (1989) Mechanisms of prey selection by stream-dwelling stoneflies. *Ecology* **70**, 1203–18.

Persson L. (1990) Predicting ontogenetic niche shifts in the field: what can be gained by foraging theory? In *Behavioural Mechanisms of Food Selection* (ed. by R.N. Hughes), *NATO ASI series, vol. G 20,* pp. 303–21. Springer Verlag, Berlin.

Persson L. & Greenberg L.A. (1990) Optimal foraging and habitat shift in perch (*Perca fluviatilis*) in a resource gradient. *Ecology* **71**, 1699–1713.

Philips-Conroy J.E. (1986) Baboons, diet and disease: food plant selection and schistosomiasis. In *Current Perspectives in Primate Social Dynamics* (ed. by D.M. Taub & F.A. King), pp. 287–304. Von Nostrand Reinhold, New York.

Pulliam H.R. (1985) Foraging efficiency, resource partitioning, and the coexistence of sparrow species. *Ecology* **66**, 1829–36.

Pulliam H.R. & Caraco T. (1984) Living in groups: is there an optimal group size? In *Behavioural Ecology: An Evolutionary Approach*, 2nd edn. (ed. by J.R. Krebs & N.B. Davies), pp. 127–47. Sinauer Assoc., Sunderland, MA.

Pyke G.H. (1984) Optimal foraging theory: a critical review. *Ann. Rev. Ecol. Syst.* **15**, 523–75.

Ramcharan C.W. & Sprules W.G. (1991) Predator-induced behavioral defenses and its ecological consequences for two calanoid copepods. *Oecologia* **86**, 276–86.

Rohwer S. & Ewald P.W. (1981) The cost of dominance and advantage of subordination in a badge signalling system. *Evolution* **35**, 441–54.

Savoleinen R. (1991) Interference by wood ant influences size selection and retrieval rate of prey by *Formica fusca. Behav. Ecol. Sociobiol.* **28**, 1–7.

Schluter D. & Grant P.R. (1982) The distribution of *Geospiza difficilis* in relation to *G. fulginosa* in the Galapagos islands: tests of three hypotheses. *Evolution* **36**, 1213–26.

Schoener T.W. (1982) The controversy over interspecific competition. *Amer. Sci.* **70**, 586–95.

Schoener T.W. (1983) Field experiments on interspecific competition. *Am. Nat.* **122**, 240–85.

Schoener T.W. (1986) Resource partitioning. In *Community Ecology. Patterns and Process* (ed. by J. Kikkawa & D.J. Anderson), pp. 91–126. Blackwell Scientific Publ., Palo Alto, CA.

Settle W.H. & Wilson L.T. (1990) Behavioural factors affecting differential parasitism by *Anagrus epos* (Hymenoptera: Mymaridae), of two species of Erythoneurian leafhoppers (Homoptera: Cicadellidae). *J. Anim. Ecol.* **59**, 877–89.

Sih A. (1987) Predator and prey lifestyles: an evolutionary and ecological overview. In *Predation: Direct and Indirect Impacts on Aquatic Communities* (ed. by W.C. Kerfoot & A. Sih), pp. 203–24. University of New England Press, Hanover, N.H.

Sih A. (1992) Prey uncertainty and the balancing of antipredator and feeding needs. *Am. Nat.* 139, 1052–69.

Sih A., Crowley P.H., McPeek M.A., Petranka J.W. & Strohmeier K. (1985) Predation, competition and prey communities: a review of field experiments. *Ann. Rev. Ecol. Syst.* 16, 269–311.

Sih A. & Moore R.D. (1990) Interacting effects of predator and prey behavior in determining diets. In *Behavioural Mechanisms of Food Selection* (ed. by R.N. Hughes), *NATO ASI series, vol. G 20,* pp. 771–96. Springer Verlag, Berlin.

Smith J.N.M., Grant P.R., Grant B.R., Abbott I. & Abbott L.K. (1978) Seasonal variation in feeding habits of Darwin's ground finches. *Ecology* 59, 1137–50.

Soluk D.A. & Collins N.C. (1988) Synergistic interactions between fish and stoneflies: facilitation and interference among stream predators. *Oikos* 52, 94–100.

Speirs D.C., Sherratt T.N. & Hubbard S.F. (1991) Parasitoid diets: does superparasitism pay? *Trends Ecol. Evol.* 6, 22–5.

Spiller D.A. & Schoener T.W. (1990) Lizards reduce food consumption by spiders: mechanisms and consequences. *Oecologia* 83, 150–61.

Stephens D.W. & Krebs J.R. (1986) *Foraging Theory.* Princeton University Press, Princeton, N.J.

Struhsaker T.T. & Leakey M. (1990) Prey selectivity by crowned hawk eagles on monkeys in the Kibale Forest, Uganda. *Behav. Ecol. Sociobiol.* 26, 435–43.

Thompson D.B.A. (1986) The economics of kleptoparasitism: optimal foraging, host and prey selection by gulls. *Anim. Behav.* 34, 1189–205.

Thompson D.B.A. & Barnard C.J. (1984) Prey selection by plovers: optimal foraging in mixed-species groups. *Anim. Behav.* 32, 554–63.

Valone T.J. & Lima S.L. (1987) Carrying food items to cover for consumption: the behavior of ten bird species feeding under the risk of predation. *Oecologia* 71, 286–94.

Vance R.R. & Schmitt R.J. (1979) The effect of predator-avoidance behavior of the sea urchin, *Centrostephanus coronatus* on the breadth of its diet. *Oecologia* 44, 21–5.

Vollrath F. (1984) Kleptobiotic interactions in invertebrates. In *Producers and Scroungers. Strategies of Exploitation and Parasitism* (ed. by C.J. Barnard), pp. 61–94. Chapman & Hall, New York.

Werner E.E. (1977) Species packing and niche complementarity in three sunfishes. *Am. Nat.* 111, 553–78.

Werner E.E., Gilliam J.F., Hall D.J. & Mittelbach G.G. (1983) An experimental test of the effects of predation risk on habitat use in fish. *Ecology* 64, 1540–8.

Werner E.E. & Hall D.J. (1976) Niche shifts in sunfishes: experimental evidence and significance. *Science* 191, 404–6.

Werner E.E. & Hall D.J. (1977) Competition and habitat shift in two sunfishes (Centrarchidae). *Ecology* 58, 869–76.

Werner E.E. & Hall D.J. (1979) Foraging efficiency and habitat shift in competing sunfishes. *Ecology* 60, 256–64.

Willis E.O. (1968) Studies of the behavior of Innulated and Salvin's antbirds. *Condor* 70, 128–48.

Wright D.I. & O'Brien W.J. (1984) The development and field test of a tactical model of the planktivorous feeding of white crappie (*Pomoxis annularis*). *Ecol. Monogr.* 54, 65–98.

Index

Page numbers in *italics* indicate figures or tables.

213